Advancements in Indoor Environmental Quality and Health

Edited by Piero Bevilacqua

Published in London, United Kingdom

Advancements in Indoor Environmental Quality and Health
http://dx.doi.org/10.5772/intechopen.111180
Edited by Piero Bevilacqua

Contributors
Alexandra Flores, Ali Bahadori-Jahromi, Changfeng Fu, Daniela Cirone, Eddie Yin-Kwee Ng, Forough Farhadi, Ijaz Fazil Syed Ahmed Kabir, Jie Deng, Mehdi Khakzand, Mohan Kumar Gajendran, Mukhtar Maigari, Natale Arcuri, Ravi Chandra Thota, Roberto Bruno, Sabrina Romano, Saeid Chahardoli, Sandra Costanzo, Sudhakar Vadivelu

First published in London, United Kingdom, 2024 by IntechOpen
IntechOpen is the global imprint of INTECHOPEN LIMITED, registered in England and Wales, registration number: 11086078, 167-169 Great Portland Street, London, W1W 5PF, United Kingdom

British Library Cataloguing-in-Publication Data
A catalogue record for this book is available from the British Library

Additional hard and PDF copies can be obtained from orders@intechopen.com

Advancements in Indoor Environmental Quality and Health
Edited by Piero Bevilacqua
p. cm.

This title is part of the Civil Engineering Book Series, Volume 7
Topic: Construction Engineering
Series Editor: Assed Haddad
Topic Editor: Amin Akhnoukh

Print ISBN 978-0-85014-018-7
Online ISBN 978-0-85014-019-4
eBook (PDF) ISBN 978-0-85014-020-0
ISSN 3029-0287

For EU product safety concerns:
IN TECH d.o.o., Prolaz Marije Krucifikse Kozulić 3, 51000 Rijeka, Croatia,
info@intechopen.com or visit our website at intechopen.com.

IntechOpen Book Series
Civil Engineering
Volume 7

Aims and Scope of the Series

Civil engineering is a traditional field of engineering from which most other branches of engineering have evolved. It comprises traditional sub-areas like transportation, structures, construction, geotechnics, water resources, and building materials. It also encompasses sustainability, risk, environment, and other concepts at its core. Historically, developments in civil engineering included traditional aspects of architecture and urban planning as well as practical applications from the construction industry. Most recently, many elements evolved from other fields of knowledge and topics like simulation, optimization, and decision science have been researched and applied to increase and evolve concepts and applications in this field. Civil engineering has evolved in the last years due to the demands of society in terms of the quality of its products, modern applications, official requirements, and cost and schedule restrictions. This series addresses real-life problems and applications of civil engineering and presents recent, cutting-edge research as well as traditional knowledge along with real-world examples of developments in the field.

Meet the Series Editor

Professor Assed N. Haddad is a Civil Engineer with a degree from the Federal University of Rio de Janeiro (UFRJ) earned in 1986, as well as a Juris Doctor degree from the Fluminense University Center earned in 1993, and a Master's degree in Civil Engineering from the Fluminense Federal University (UFF) obtained in 1992. He completed his Ph.D. in Production Engineering from COPPE / Federal University of Rio de Janeiro in 1996. Professor Haddad's academic pursuits have taken him to postdoctoral stays at the University of Florida, USA in 2006; at the Universitat Politècnica de Catalunya, Spain in 2010; and at the University of New South Wales Sydney, Australia in 2019. Currently, he serves as a Full Professor at the Federal University of Rio de Janeiro. He has held visiting professorships at various institutions including the University of Florida, Universitat Politècnica de Catalunya, Universitat Rovira i Virgili, and Western Sydney University. His research expertise encompasses Civil, Environmental, and Production Engineering, with a primary focus on the following topics: Construction Engineering and Management, Risk Management, and Life Cycle Assessment. He has been the recipient of research grants from the State of Rio de Janeiro, Brazil: CNE FAPERJ from 2019 to 2022 and from 2023 to 2025. Additionally, his research grants obtained from the Brazilian Government CNP since 2012 last to this date. Professor Haddad has been involved in several academic endeavors, being the Guest Editor of the International Journal of Construction Management; MDPI's Sustainability, Energies, and Infrastructures; Associate Editor at Frontiers in Built Environment / Sustainable Design and Construction; Guest Editor at Frontiers in Built Environment / Construction Management; and Academic Editor of the Journal of Engineering, Civil Engineering Section of Hindawi. He is currently a Professor of the Environmental Engineering Program at UFRJ and the Civil Engineering Program at UFF.

Meet the Volume Editor

Piero Bevilacqua is a researcher at the Department of Mechanical, Energy and Management Engineering, University of Calabria, Italy. He received his Ph.D. in Mechanical Engineering in 2014. He has co-authored several journal papers and two book chapters. He has actively participated in numerous research projects in energy efficiency. His main areas of interest include passive systems for the building envelope, green roofs, innovative photovoltaic systems, thermal comfort of indoor spaces and Indoor Environmental Quality (IEQ), nearly zero-emission building (NZEB) in the Mediterranean, innovative solar-assisted air-conditioning plants, integrated thermal storage systems, solar cooling, thermal properties of building materials, and renewable cogeneration systems.

Contents

Preface

It is evident how current urban society is structured in a way that forces humans to spend a significant portion of their lives in confined spaces, whether it be a residence or workplace. In recent years, there has been a growing interest in improving Indoor Environmental Quality (IEQ), driven by an increased awareness of the impact of this aspect on people's health and productivity.

Factors influencing IEQ and occupant satisfaction can be categorized into physical and non-physical factors. Physical factors encompass four aspects: thermal comfort, indoor air quality, lighting, and acoustic environment, which can be assessed using measurable parameters. Non-physical factors generally refer to internal qualities that are challenging to measure with instruments, including space, privacy, furnishings, cleanliness, facilities, and the view.

Considering that the topic is highly complicated, it is imperative to approach it from an interdisciplinary standpoint. Aspects to be taken into consideration undoubtedly include the continuous evolution towards increasingly higher comfort standards and the monitoring of an ever-growing number of parameters using increasingly sophisticated measuring tools. What is hoped for in this fascinating field of research is the definition of a global index or a single well-being scale that surpasses the narrow fields in which we currently operate.

This book gathers a variety of research and studies that present the latest knowledge linked to IEQ and synthesizes diverse studies in IEQ engineering. Starting with reviews of crucial factors such as thermal comfort and air quality, the book delves into technological applications like machine learning for IEQ assessment. Chapters explore the intersection of IEQ, sustainability, and occupant wellbeing, laying the groundwork for future engineering practices. The compendium highlights the evolution of IEQ standards and protocols, emphasizing the role of technology in shaping healthier indoor environments. It also explores various aspects of ventilation strategies in buildings, emphasizing their role in mitigating airborne disease transmission. The book concludes by emphasizing the integration of machine learning techniques into IEQ assessment, demonstrating their potential in optimizing building performance and occupant satisfaction.

Drawing insights from a spectrum of technical works, this collection meticulously dissects the multifaceted relationship between Indoor Air Quality (IAQ), IEQ, and the evolving technological paradigm. Prominent studies underscore the potential of emerging technologies to surmount existing challenges, providing a trajectory of sustainable and healthier indoor spaces.

Piero Bevilacqua
Department of Mechanical,
Energy and Management Engineering,
University of Calabria,
Rende, Italy

Chapter 1

Monitoring Indoor Air Quality in Buildings: An Overview of Measuring Devices and Main Challenges for a Correct Operation

Daniela Cirone, Sabrina Romano, Roberto Bruno and Natale Arcuri

Abstract

This research delves into the foundational elements of thermal comfort, crucial alongside visual comfort, acoustic comfort, and air quality for ensuring the quality and sustainability of living environments. With increasing recognition of thermal comfort's implications across various human activities, from energy management for efficiency to considerations of environmental impact and economy, its comprehensive understanding is paramount. The study scrutinizes the prevailing methodology of evaluating comfort *via* true thermal sensation rating, dissecting the involved variables and their relative significance in determining comfort levels. Following this analysis and parameter definition, a comparative assessment of diverse sensors was conducted to gauge measurement accuracy concerning key variables of interest, thereby identifying the most suitable sensor for real-world applications. Conducted at the ENEA research center in Rome, the study executed an experimental setup within one of the center's offices. Subsequently, reflections were made on the feasibility of providing indoor comfort indications amidst variable data availability, exploring potential simplifications and approximations to streamline comfort index evaluations.

Keywords: thermal comfort, PMV, humidex, sperimental data, *EN ISO 7726*

1. Introduction

Indoor air quality (IAQ) monitoring within buildings is a critical endeavor, impacting the health, productivity, and comfort of occupants. This article provides a comprehensive overview of the various measuring devices employed for monitoring IAQ, along with an examination of the primary challenges encountered in ensuring their effective operation. Additionally, we present a detailed case study where both commercial and experimental sensors are analyzed, showcasing their respective characteristics and performance. With a growing emphasis on healthy indoor environments, understanding the landscape of IAQ measurement tools and addressing

IntechOpen

associated challenges, as evidenced by real-world applications, is paramount for maintaining optimal indoor air quality standards [1].

EN ISO 7726 [2] standard is a document in the series of international standards that specifies the minimum requirements of instruments for measuring physical quantities describing an indoor environment, providing recommended measurement approaches considering pragmatic options and performance of the available measuring instruments. The specifications contained in this standard are divided into two classes, according to the magnitude of the monitored quantities. The first is defined as "Type C" describing specifications and methods for measurements carried out in moderate environments, for instance how to record thermo-hygrometric comfort conditions following the indices of EN ISO 7730 [3]; the other class is "Type S" containing specifications and methods for environments exposed to thermal stress by the indications of EN ISO 7933 [4] for hot environments and EN ISO 11079 [5] for cold environments. The suitable parameters for determining thermo-hygrometric comfort or thermal stress indexes can be divided into basic and derived physical quantities. The firsts are used to define comfort or thermal stress indices such as air temperature (in K or °C), air velocity (in m/s), mean radiant temperature (in K or °C), and absolute humidity expressed as the partial pressure of water vapor (Pa). The derived physical quantities are evaluated as a function of the sensor features and are employed for determining empirical indexes related to thermo-hygrometric comfort conditions or thermal stress avoiding deterministic methods that rely on the thermal exchanges between the human body and surroundings. Methods of measuring the environmental physical properties must take into account characteristics changeable with position and time, for instance, a human body considered sitting or standing.

2. Physical comfort variables

Physical comfort is a multifaceted construct encompassing various environmental factors that influence an individual's subjective experience of well-being and satisfaction. These factors, referred to as physical comfort variables, play a pivotal role in shaping the overall comfort level within indoor environments. Understanding and effectively managing these variables is essential for optimizing occupant comfort, productivity, and overall quality of life. Key physical comfort variables include air temperature, relative humidity, air velocity, and mean radiant temperature, among others. Each of these variables interacts dynamically to create a thermal environment that can either promote or hinder comfort. Additionally, factors such as clothing insulation, metabolic rate, and individual preferences further modulate the perception of comfort, highlighting the intricate interplay between environmental conditions and personal characteristics.

2.1 Air temperature

Among the quantities involved in microclimate evaluations, air temperature is often used to measure the dry-bulb temperature, most easily measurable in light of the several consolidated technologies available, both by traditional or innovative probes. Traditional technologies include liquid or solid expansion thermometers, resistance thermometers, thermocouples, and platinum resistance thermometers. Innovative technologies involve the employment of optic fibers and infrared thermometers. However, for this sensor typology, it is required to shield it adequately to prevent a measurement affected by thermal radiation emitted from surrounding bodies at different temperatures than the air temperature. In order to give a quantitative indication, a not shielded sensor or inappropriately shielded installed inside an indoor environment with an average radiant temperature 10°C higher than the air temperature, errors of even more than 1°C could be detected. The EN ISO 7726 [2] standard establishes the required features of sensors employed for air temperature measurements. For class C measurements, instruments with a maximum desirable error of less than ±0.2°C and required ±0.5°C are recommended, other parameters instead are given for class S. Regarding the response time, the standard states that this should be as small as possible for both classes, but without specifying a limiting value. However, it is recommended that the measurement duration should be at least 1.5 times the instrument response time, because a high response time may be critical for the monitoring of parameters in non-steady-state conditions.

2.2 Mean radiant temperature

Different solutions are available for the measurement of the mean radiant temperature in a confined space, therefore also different measuring instruments can be employed for this purpose as indicated in the standard EN ISO 7726 [2]. The mean radiant temperature assumes a significant role in the evaluation of thermal environment [6] features because it represents the uniform temperature of a fictitious black cavity in which a subject would exchange the same amount of radiant energy as that could be exchanged in the real non-uniform environment (the correspondent symbol is denoted by *tr*). The standard outlines three techniques for measurement: using a globo-thermometer, employing a two-sphere radiometer, and utilizing a constant air temperature sensor. Additionally, it introduces two calculation approaches grounded in the view factors between the cavity's surface and the radiant temperatures of the plane.

For this parameter, the most widespread instrument is undoubtedly the globo-thermometer because it is also the cheapest sensor, which, however, suffers from different drawbacks, such as the high response time (normally more than 20 minutes), which leads to evident issues when numerous measurements are requested for a limited period, and to overestimation due to the perfect spherical shape that does not consider the radiant contributions of horizontal surfaces adequately. Moreover, in moderate environments, the globe-thermometer does not allow calculation of the radiant temperature asymmetry, indispensable to determine thermo-hygrometric comfort conditions in large glazed environments (panes have a surface temperature much lower than opaque surfaces). The globe-thermometer is made of a very thin and opaque black metal sphere with an assigned diameter of 15 cm whose emission coefficient is about 0.95. A temperature sensor is instead placed inside the sphere so that the globe temperature (that after a proper time reaches the thermal equilibrium with the cavity) is determined by including the effects of the body's radiative and convective exchanges. Once the convective share is determined through temperature and air velocity, it is possible to extrapolate the average radiant temperature of the environment. In particular, the mean radiant temperature is evaluated using the equations introduced in Annex B of EN ISO 7726 [2]. The following empirical equation, for instance, is used assuming natural convection and when the globe temperature *tg* refers to a standard instrument with a diameter of 0.15 cm.

$$t_r = \left[\left(t_g + 273\right)^4 + 0.4 \cdot 10^8 \cdot \left|t_g - t_a\right|^{1/4} \cdot \left(t_g - t_a\right) \right]^{1/4} - 273 \qquad (1)$$

with *ta* is the air temperature. Another issue related to measurements carried out by a globe-thermometer is due to the measurement conducted at a single point that is not representative of the entire radiative field relative to the subject.

Regarding the available theoretical models, the view-factor method is more accurate and it determines the average radiant temperature starting from the measurement of the surface temperatures for each wall inside the cavity, in turn detected by contact or distance methods. Anyway, the estimation of the view factor between the person and the surrounding surfaces is required and its calculation is quite difficult. Indeed, the view factor is a function of shape, size, and relative positions concerning the person and the surface. Compared with the other methods, the view factor method benefits from better accuracy in determining radiant contributions and the ability to estimate asymmetric conditions as well. However, problems such as calculating the view factors in complex spatial geometries and measuring the mean surface temperature arise.

An alternative measurement can be conducted through a radiometer, which consists of a flat thermal element with high emissivity measuring the flux of the incident radiant energy by a thermopile, representing the sensing element, as a function of absorbed heat. The sensor temperature *Ts* is then directly related to the radiant plane temperature Tpr according to a balance equation. The average radiant temperature can then be evaluated by measuring the radiant plane temperature in the six space directions and the projected area factors. Radiant asymmetry can also be assessed by using the same approach.

2.3 The relative air velocity

The measurement of air velocity is quite complex, both because of its rapid temporal fluctuation and its vector nature with great variability in intensity and direction. Omnidirectional probes allow measurement of the modulus of velocity regardless of their placement, while one-way probes must be placed orthogonally to the flow and must be more than one in number to allow measurement of individual velocity components. The velocity modulus can be determined using a single omnidirectional probe, such as a hot bulb probe, or in the case where the flow is unidirectional with a single one-way probe, such as a hot-wire anemometer. The Annex E of the standard EN ISO 7726 [2] considers the use of wire/hot film anemometers, whirlwind anemometers, and ultrasonic anemometers in the measurement of wellbeing, but the direction sensitivity and response time of the instrument should be evaluated in advance.

In wire and hot-film anemometers, velocity monitoring is based on measuring the exchange of thermal energy between the sensing element and the surrounding air. The anemometer consists of a solid body, which in the case of wire sensors has the shape of a cylinder or with different shapes for film sensors, such as that of a sphere. The sensor body is heated electrically by a higher temperature than the surrounding air, giving up thermal energy mainly by convection. The heating power, the temperature of the element, and the air temperature allow for calculating the air velocity by the convective heat transfer coefficient. Hot-sphere sensors are isotropic, namely, they show equal sensitivity to flows from all spatial directions, but the significant mass makes these sensors slower in response and unsuitable for tracking the rapid changes in a turbulent airflow. Hot-wire sensors are directional and therefore their sensitivity varies significantly between the perpendicular plane to the wire and the parallel direction to the wire anyway they are fast in response. Whirlwind anemometers consist of a small impeller with blades or cups suspended in the flowing stream, with their axis of rotation coaxial or perpendicular to the flow direction. They are therefore unidirectional sensors and it is needed to know in advance the velocity vector track, moreover, they are typically less sensitive. Ultrasonic anemometers use special phenomenology related to the propagation of ultrasonic waves through a moving fluid for air velocity detection. Based on the physical principle used, they can be divided into transit time meters and meters and the Doppler effect. In both, the propagation of pressure waves at frequencies higher than those audible by the human ear and propagating within the fluid current is exploited. Hot-wire anemometer normally produces negligible errors and is indicated in the context of PMV calculation. If the velocity is very low and no dominant direction of flow is identified, the effect of air velocity is inevitably small. If, on the contrary, there is appreciable air flow, its direction becomes obvious and the sensor can be oriented accordingly to maximize its sensitivity. More accurate velocity meters such as Pitot tubes can be used to periodically check anemometers, and the instrument can be calibrated at least every 2 years.

2.4 Relative humidity

The humidity content in air has a significant influence on an individual's thermo-hygrometric balance, especially on evaporative transmissive share. High air humidity drastically reduces the evaporation of sweat, and in the summer regime, this rate can account for as much as 70–80% of the entire heat transfer. In thermo-hygrometric air conditioning problems, atmospheric air is considered a binary mixture made of dry air and water vapor called "moist air." In turn, dry air is a mixture of gases consisting mainly of nitrogen and oxygen, therefore this component is approximated as a single component with ideal gas behavior with invariable composition. So, moist air is considered as a mixture of perfect gases, however with different properties than mixtures of perfect gases because the water vapor can be subjected to phase exchange processes, altering the composition of the mixture. Moist air can contain an amount of vapor varying from zero (dry air) to a maximum value (saturated humid air) that depends on the temperature and pressure levels of the mixture. At any given temperature, air cannot contain more than a certain amount of water vapor; above that amount, water vapor condenses forming a liquid phase. In these conditions, the saturation point at a precise temperature and pressure is reached. As the temperature of the air increases, the maximum amount of water vapor it can contain also increases, consequently humid air mixtures at high temperatures contain more water vapor than the same mixtures at low temperatures. To quantify the magnitude of the water vapor within the mixture, the concept of absolute humidity (or humidity ratio) is defined, to consider how much the most air is distant from the saturation point, relative humidity instead is introduced.

Absolute humidity refers to the quantity of water vapor, measured in grams, present within one cubic meter of air under specific temperature and pressure conditions. As noted earlier, absolute humidity typically rises as temperature increases. To better highlight its physical meaning, it is expressed in kg of vapor (kg^v) per kg of dry air (kg^a) [kg^v/kg^a or g^v/kg^a, the last more used considering that the quantity of water vapor is very smaller than the mass of dry air]. Absolute humidity is measured directly, using hygrometers with lithium chloride salts, hair, and dew point. Its value varies between zero (complete dry air) and an infinity value, namely when the mixture is composed exclusively of water vapor.

Relative humidity (ϕ) is the ratio of the mass m^v of water vapor present in a certain volume V of moist air to the mass m^s of vapor containable under saturated conditions

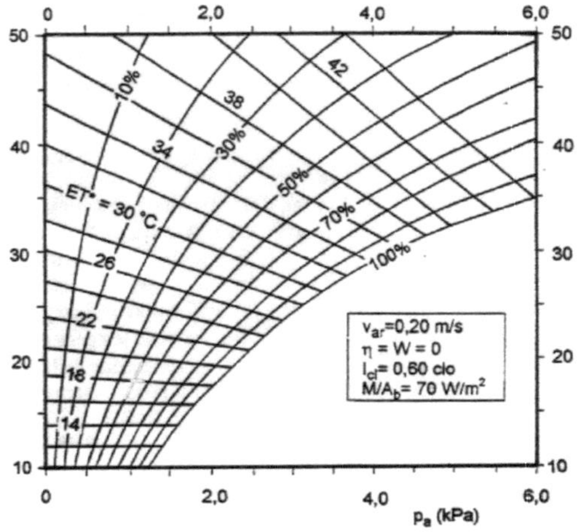

Figure 1.
Psychrometric diagram with ET curves, constant for M/ab = 1.2 met, Icl = 0.60 clo, var. = 0.20 m/s.*

at the same temperature in the same volume V of moist air. It is also expressed as the ratio of the vapor pressure to the saturation pressure of vapor at the temperature *ta* of the moist air mixture.

$$\varphi = \frac{p_v}{|p_s|_{ta}} \tag{2}$$

Relative humidity is the most widely used hygrometric parameter for practical reasons. Since the capacity of air to hold water vapor increases with temperature, it follows that, leaving the pressure and the amount of water vapor contained in a given volume of air unchanged, relative humidity decreases as temperature increases and *vice versa*. This is because as temperature increases, more water vapor is required to reach saturation. Relative humidity is measured indirectly with a psychrometer (**Figure 1**). This instrument consists of two temperature probes: the first probe is in direct contact with the air, measuring the "dry bulb temperature," the second probe is wrapped in moistened gauze and equipped with a ventilation system that facilitates the evaporation of water, measuring the "wet bulb temperature." Due to the cooling effect of evaporation (the water absorbs heat from the air to vaporize) the wet probe measures a lower temperature and the humidity of the air increases in light of the greater vapor quantity. By comparing the two temperatures using appropriate diagrams (called psychrometric) or algorithms, the relative humidity value is detected.

Relative hygrometers, such as hair hygrometers, capacitive and resistive electrical hygrometers can measure relative humidity directly. The principles of measurement (mechanical, electrical, resonance, etc.), production technologies (thin-film, thick-film, solid-state, etc.), and materials used (polymeric, ceramic, etc.) are the most varied allowing this parameter to be measured over a wide range (10–100%), with an uncertainty typically of 2–3% and at best less than 0.5%.

2.5 The operating temperature, equivalent temperature, and effective temperature

To characterize a confined thermal environment using fewer parameters, and avoiding measuring the mean radiant temperature (the monitoring of which is laborious), built-in parameters have been introduced. The three most important are operating temperature (*to*), equivalent temperature (*teq*), and effective temperature (*ET**). The integrated parameters combine the influence of individual parameters on energy dissipation in the following way:

to integrate the effect of ta + tr
teq integrated effect of ta + tr + va
*ET** integrated effect of ta + tr + pa

The dependence of PMV on air temperature and mean radiant temperature, which determines convective and radiative exchanges, respectively, is traced back to that from a single variable, a linear combination of these two temperatures and is named operating temperature *too*. The operating temperature is rigorously defined as:

$$t_0 = \frac{h_r \cdot t_r + h_c \cdot t_a}{h_r + h_c} \tag{3}$$

where
Ta is the air temperature, in [°C],
Tr is mean radiant temperature, in [°C],
hc is the convective heat transfer coefficient, in [$W{\cdot}m^{-2}{\cdot}K^{-1}$],
hr is the radiative heat transfer coefficient, in [$W{\cdot}m^{-2}{\cdot}K^{-1}$].

The operating temperature corresponds to the temperature actually felt by people that takes into account the main heat exchange mechanisms affecting the human body on the feeling of well-being. Alternatively, *t0* can be calculated, in a simplified way, by the equation:

$$t_0 = A \cdot t_a + (1 - A) \cdot t_r \tag{4}$$

where A, listed in **Table 1**, represents a parameter related to the relative air velocity. It can be appreciated that the higher the air velocity, that is, convective heat transfer coefficient, the greater the weight of air temperature and the lower the weight of mean radiant temperature.

Another simplified formula can be adopted:

$$t_0 = \frac{t_r + t_a}{2} \tag{5}$$

but only when precise conditions are respected. In particular, the human metabolism (related to the activity and measured in *met* following the Fanger theory, see EN ISO 7730) must range between 1.0 and 1.3 met, air velocity lower than $0.2 m{\cdot}s^{-1}$, absence of direct solar radiation and difference between air temperature and mean radiant temperature below 4°C.

V_{ar}	$V_{ar} < 0.2$	$0.2 < V_{ar} < 0.6$	$0.6 < V_{ar} < 1.0$
A	0.5	0.6	0.7

Note:
$V_{ar} = V_a + 0.0052 \cdot \left(\frac{M}{Ab} - 58.15\right)$

Table 1.
Values of parameter a as a function of relative velocity expressed in $m{\cdot}s^{-1}$.

Starting from the consideration that relative humidity has a negligible effect on thermal comfort conditions, Dufton in 1929 had the idea of combining the other three physical parameters (ta, tr, and va) into a single index defined equivalent temperature *teq*, and defined as the temperature of the thermally uniform fictitious environment (ta = tr) with tinned air in which a person would exchange the same dry heat power as in the real environment. In residential environments, the use of the operating temperature is more recommended, considering that va generally takes on very low values. In different cases where va takes on values that are not considered negligible, such as analysis inside vehicles, the equivalent temperature comes into play. One of the last equations suggested for calculating teq is that of Madsen (1979):

When $v_a > 10$m/s

$$t_{eq} = 0.55 \cdot t_a + 0.45 \cdot t_r + (0.24 - 0.75 \cdot \sqrt{v_a}) \cdot (36.5 - t_a)/(1 + I_{cl} \quad (6)$$

$$\text{When } v_a \leq 10 \text{ m/s } t_{eq} = 0.5 \cdot (t_a + t_r) \quad (7)$$

The ET* index, known as new effective temperature, considers the temperature of a fictitious environment at a uniform temperature (ta = tr) and with a hygrometric degree of 0.5 in which a human body would exchange by convection (C), radiation (R), and evaporation (E) the same amount of heat that could exchange in the real environment, at parity of skin temperature *tsk* and the same percentage of wet skin *w* as in the real environment. The calculation of ET* is carried out by using a thermoregulation model, and knowing the values of the six variables on which the global thermal discomfort depends, and the values of tsk, w, and (R + C + E) are calculated. Then, the temperature of the uniform fictitious room ta* = tr*, at $\phi = 0.5$ in which, with real tsk and w, the subject would exchange a heat output (R* + C* + E*) = (R + C + E), is determined. The ta* thus calculated is precisely ET*. A psychrometric diagram (**Figure 1**) allows for determining ET*, as a function of the partial pressures on the abscissae and the operating temperatures on the ordinates, by setting the values of air velocity, metabolic rate, and clothes thermal resistance.

To summarize, **Table 2** lists the main characteristics of measuring instruments according to the EN ISO 7726 prescriptions.

3. Evaluation of measurements and discomfort

Currently, there is no specific standard in the European landscape that defines the criteria to be followed when conducting monitoring aimed at assessing thermal comfort in an existing building [7]. Some guidance is provided by the standard EN ISO 7726, which indicates how methods for measuring the physical characteristics of the environment consider the parameter variation with time and space.

An environment can be considered bio-climatically homogeneous if, at any given time, air temperature, radiation, air velocity, and humidity can be considered uniform around the subject. This condition is frequently met for temperature, air velocity, and humidity, but more rarely in the case of radiation [8]. In monitoring, it is aimed at determining the comfort level, measurements should be made at several positions representative of the average and/or worst conditions around the subject. The determination of measurement locations is done to capture the heterogeneity of thermal environments. In a homogeneous indoor space, a single measurement could be made at a conventional point, such as the center of the room, or multiple measurements that correspond to the spatial gradient of thermo-hygrometric parameters.

Quantity	Symbol	Class C (comfort)			Class S (thermal stress)		
		Measuring range	**Accuracy**	**Response time (90%)**	**Measuring range**	**Accuracy**	**Response time (90%)**
Air temperature	t_a	10–40°C	*Required*: ±0.5 °C *Desirable*: ±0.2 °C These levels shall be guaranteed at least for a deviation $\|t_r\text{-}t_a\|$ = 10°C	The shortest possible. Value to be specified as characteristic of the measuring instrument.	−40°C to +120°C	*Required*: −40–0°C: ±(0.5 + 0.01$\|t_a\|$)°C > 0–50°C: ±0.5°C > 50–120°C: ±[0.5 + 0.04 (t_a−50)]°C *Desirable*: required accuracy/2 These levels shall be guaranteed at least for a deviation$\|t_r−t_a\|$ = 20°C	The shortest possible. Value to be specified as characteristic of the measuring instrument.
Mean radiant temperature	t_r	10–40°C	*Required*: ±2°C *Desirable*: ±0.2°C These levels are difficult or even impossible to achieve in certain cases with the equipment normally available. When they cannot be achieved, indicate the actual measuring precision.	The shortest possible. Value to be specified as characteristic of the measuring instrument.	40°C to +150°C	*Required*: −40–0°C: ±(5 + 0.02$\|t_a\|$)°C >0–50°C: ±5°C >50–150°C: ±[5 + 0.08 (t_r−50)]°C *Desirable*: −40–0°C: ±(0.5 + 0.01$\|t_r\|$) °C >0–50°C: ±5°C >50–120°C: ±[0.5 + 0.04 (t_r−50)]°C	The shortest possible. Value to be specified as characteristic of the measuring instrument.
Air velocity	V_a	0.05 m/s to 1m/s	*Required*: ±(0.05 + 0.05 V_a) m/s *Desirable*: ±(0.02 + 0.07 V_a) m/s These levels shall be guaranteed whatever the direction of flow within a solid angle.	Required: 0.5 s Desirable: 0.2 s	0.2 m/s to 20 m/s	*Required*: ±(0.1 + 0.05 V_a)m/s *Desirable*: ±(0.05 + 0.05 V_a)m/s These levels shall be guaranteed whatever the direction of flow within a solid angle.	The shortest possible. Value to be specified as characteristic of the measuring instrument. For measuring the degree of turbulence, a small response time is needed.
Absolute Humidity as partial pressure of water vapor	Pa	0.5 kPa to 3.0 kPa	±0.15 kPa This level shall be guaranteed for a difference $\|t_r\text{-}t_a\|$of at least 10°C.	The shortest possible. Value to be specified as characteristic of the measuring instrument.	0.5 kPa to 6.0 kPa	±0.15 kPa This level shall be guaranteed for a difference $\|t_r\text{-}t_a\|$of at least 20°C.	The shortest possible. Value to be specified as characteristic of the measuring instrument.

Table 2.
Characteristics of measuring instruments EN ISO 7726.

The EN ISO 7726 [2] standard indicates at which heights the basic quantities have to be measured and the weighting coefficients to use about the class of the environment (**Table 3**). A space is said to be stationary when the physical quantities used to describe the thermal environment are time-independent. This condition is reached when the parameter fluctuations, about their time average, do not exceed the values obtained by multiplying the measurements with the factors given by EN ISO 7726 (**Table 4**) [2].

It should be noted that individual quantities used to describe the level of exposure (metabolism, heat insulation) may also be time-dependent.

Measurements should also be carried out concerning weather conditions representative of hot and cold seasons. The duration of individual monitoring should be sufficient to comprehensively represent the reference condition, also considering the response times of different probes. The results of the measurements are inevitably influenced by the methodological choices made, which are decisive to ensure the highest level of repeatability and to produce an accompanying report that contains useful information to contextualize the results.

A description of the sample environments, a description of the test conditions, the criterion used for the choice of measurement points, the duration of the measurements, the external meteorological conditions, and an accurate description of the sensor used are indicated in the following.

The conditions $-0.5 < PMV < +0.5$ and $5\% < PPD < 10\%$ represent necessary but not sufficient conditions for comfort in average comfortable environments. To attain actual comfort there must also be zero discomfort due to the unevenness of environmental variables, that is, there must be no local discomfort [9]. Along with PMV, which condensates the energy balance of the human body, the EN ISO 7730 standard

Locations of the sensors	Weighting coefficients for measurements for calculation mean values				Recommended heights (for guidance only)	
	Homogeneous environment		Heterogeneous environment		Sitting	Standing
	Class C	Class S	Class C	Class S		
Head level			1	1	1.1 m	1.7 m
Abdomen level	1	1	1	2	0.6 m	1.1 m
Ankle level			1	1	0.1 m	0.1 m

Table 3.
Measuring heights for the physical quantities of an environment.

Quantity	Class C (comfort) Factor X	Class S (thermal stress) Factor X
Air Temperature	3	4
Mean radiant temperature	2	2
Radiant temperature asymmetry	2	3
Mean air velocity	2	3
Vapor pressure	2	3

Table 4.
Criteria for a homogeneous and steady-state environment.

contains some indices to describe "local" discomfort considering local fluctuations or disconformities of microclimatic quantities, such as to induce discomfort conditions in the subject. The main causes causing local discomfort are mainly due to temperature unevenness and air velocity fluctuations, which affect heat exchange with the surroundings. For this, four main causes can be considered.

4. Investigated case study

The experimental campaign was conducted in a part of the building hosting the ENEA Casaccia Research Center in Italy. The Casaccia Research Center is ENEA's largest laboratory and facility complex and is located 25km northwest of Rome, near Bracciano Lake. The building considered is identified as Building F40 of the Department of Energy Technologies (DTE). The data were collected in an operational office located on the second floor belonging to the smart city and smart community laboratory, identified as room 105 and highlighted in the 3D model representation of **Figure 2**.

Room 105 has a net area of 17.5 m^2, dimensions 3.88 × 4.51 m and faces east with a window opening of 1.70 × 1.40 m. The room is equipped with office furniture set up for two operating desks (**Figure 3**). The room is heated and air-conditioned by a single fan coil. In the winter period, the fan coil is supplied with hot water from a district heating system, in summer, the chilled water comes from an autonomous chiller located in the building. The room is illuminated by six OSRAM Lumilux T8 LEDs 26 mm L 36 W/840 lamps not equipped with either dimmer or modulating ignition devices.

4.1 Installed sensors

Table 5 shows schematically all the sensors considered in the study.

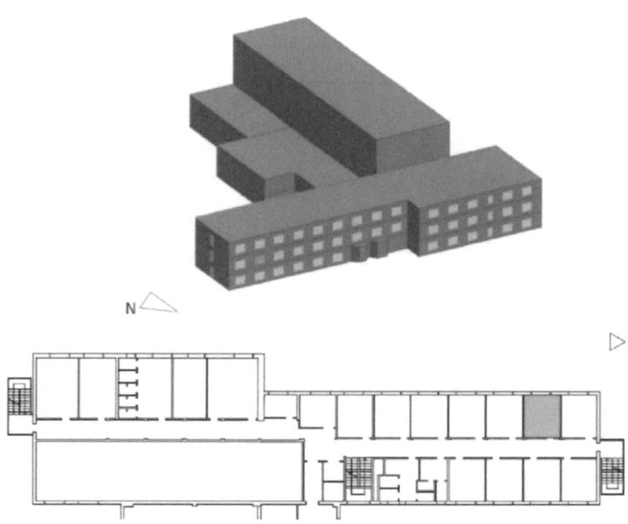

Figure 2.
3D model representation of the building and plan.

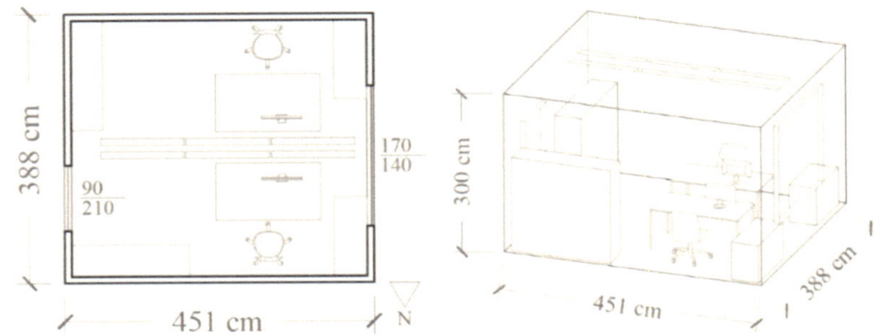

Figure 3.
Geometrical representation of the room 105.

Sesto Senso	ERS CO2	Motion Sensor FIBARO	Testo 480	CO2-display

Table 5.
Sensors installed in the case study office.

4.2 Sesto senso

Sesto Senso is a patented multisensor system by ENEA consisting of a central unit that acts as a gateway, sensors that communicate with the gateway *via* a Z-Wave protocol, and a Z-Wave people-counting sensor whose computational algorithm was developed by ENEA itself. This sensor is derived from the optimization of a prototype (**Figure 4**), and the upgrade consisted of integrating new sensor modules, particularly

Figure 4.
Sesto senso prototype.

microphones, optimizing the previous people-counting module, and integrating new hardware and software.

The sensors included in Sesto Senso are:

Fibaro "door/window" sensor 2 (**Figure 5**). In particular, two sensors were installed; one for assessing the door status and another for the window status. The detection of the opening occurs when the sensor body and the magnet are separated. Furthermore, the Fibaro door/window sensor 2 includes an integrated temperature sensor and is powered by a 3.6 V DC battery. It has dimensions of 71 × 18 × 18 mm and a temperature measurement accuracy of ±0.5°C, which is within regulatory limits.

Multisensor 6 from Aeotec (**Figure 6**), a small device for detecting motion, temperature, humidity, illuminance, UV, and vibration. Regarding accuracy concerning temperature, it does not meet the necessary limit dictated by EN ISO 7726 [2].

An ad-hoc people counting device was realized for the monitoring, with 2 IR TF mini LIDAR sensors, based on time-of-flight technology. The maximum detection range is 12 m and the acceptance angle is 2.3°. The object has dimensions 150 × 60 × 30 mm, with two holes for laser output, power supply is 12VDC and is *via* cable (**Figure 7**).

For the measurement of CO_2 concentration, no specific sensor was used, but an indirect value was derived according to an algorithm, developed by the same research organization that owns the system. The algorithm that calculates the percentage of CO_2 has as input five environmental parameters of the room: door and window status, temperature, humidity, and number of people present; even if just one of the variables is missing, the value of CO_2 is not returned.

Figure 5.
Fibaro door/window sensor 2.

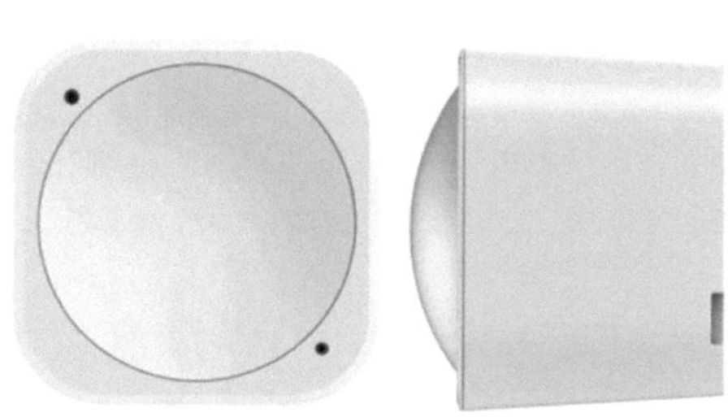

Figure 6.
Multisensor 6 Aeotec.

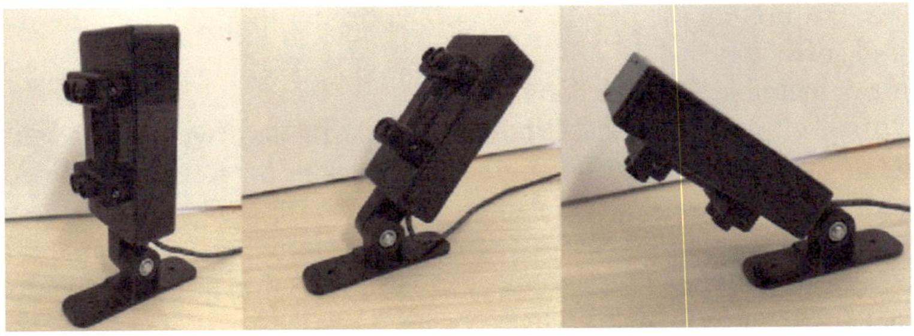

Figure 7.
Counting people Sesto senso.

The temperature value reported on the control unit is that of the center of the room to have a more indicative value of the temperature distributed in the room. The room-center temperature value is also used for the calculation of CO_2, and also for the subsequent analyses covered in this discussion, only the latter will be considered referring to the Aeotec sensor.

Most of the sensors, except the people-counting sensor, are wireless and battery-powered so that the problem of connecting them to the control unit was overcome.

Table 6 lists the main characteristics of Sesto Senso measurements.

The control unit consists of a 7″ touchscreen display whose system is developed from a Raspberry Pi 3 board. The visual interface allows the user to view all the detected parameters also through intuitive and user-friendly icons. The data-sending routine runs every two minutes and sends the last data recorded within two minutes. The touch screen, on the other hand, updates every 30 seconds (**Figure 8**).

Since Sesto Senso consists of multiple sensors, each requires specific placement according to its function. Fibaro's door/window sensor 2, in the case of the door, was placed at a height of 1.6 m, although this is irrelevant for the relief of opening and closing, while a high height could distort the temperature reading. In the case of the window, on the other hand, considering the double side and vasistas opening, it was necessary to place the sensor on the top. These problems in sensor positioning, and also the peripheral location for the working positions, led to disregarding the

Quantity	Unit	Accuracy	Range	Sampling
Temperature	[°C]	± 1 °C	−10–50°C	30 report/h
Relative humidity	[%]	± 3% (at 25°C)	20–90%	30 report/h
Illuminance	[lx]	—	0–30,000 lx	30 report/h
CO_2	[ppm]	—	—	30 report/h
Presence	N° of person	—	—	—
UV index	—	—	0–10	30 report/h

Table 6.
Technical characteristics of Sesto senso measurements.

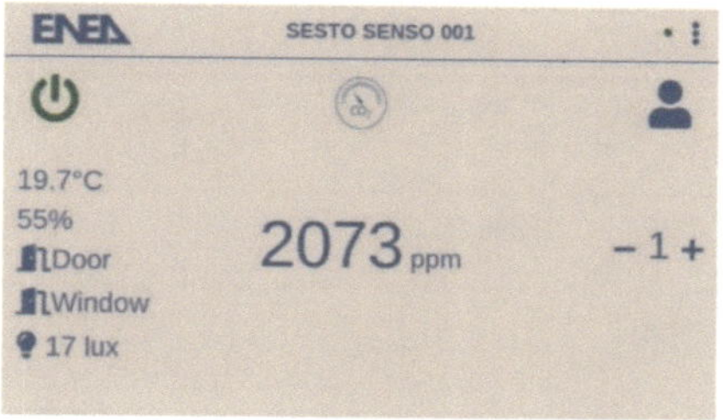

Figure 8.
Sesto senso control screen.

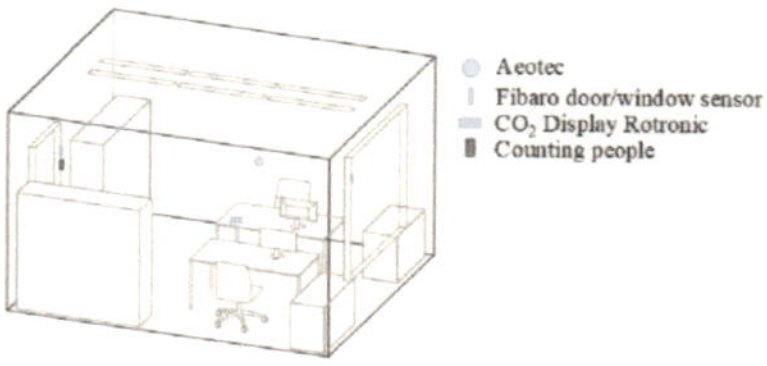

Figure 9.
Sensors position and location in room 105.

temperature values detected by this sensor and focusing only on the Aeotec one. The latter was placed on the south wall at a height of 1.60 m (**Figure 9**). Finally, the people-counting sensor was placed on the inside of the door. This sensor presented several problems in its placement, the first of which is that it detects the opening and closing of the door as the passage of a person, thus distorting the number of people inside the room and consequently also the calculation of CO_2.

4.3 ERS CO2

ERS CO2 (**Figure 10**) is a sensor for measuring indoor environment conditions, developed by ELSYS. It is designed to be wall-mounted at a height of 1.6 m. It is completely wireless and powered by two 3.6 V AA lithium batteries. The sensor box with a rectangular-shaped case, measuring 86 × 86 × 28 mm, contains sensors for measuring CO levels, air temperature, humidity, illuminance, and presence. As for temperature, it falls within the "desirable" values of the EN ISO 7726 standard (**Table 7**).

Figure 10.
ERS CO_2 sensor.

	Quantity	Unit	Accuracy	Range	Sampling
	Temperature	[°C]	±0.2 °C	−3276.5 °C–3276.5°C (Value of: 100→ 10.0°C)	12 report/h
	Relative Humidity	[%]	± 2% (a 25°C)	0–100%	12 report/h
	Presence	[on/off]	—	0–255 (number of movements counted)	—
	Illuminance	[lx]	±10 lx	0–65,535 lx	12 report/h
	CO_2	[ppm]	±50 ppm	0–10,000 ppm	12 report/h

Table 7.
Technical characteristics of ERS CO2.

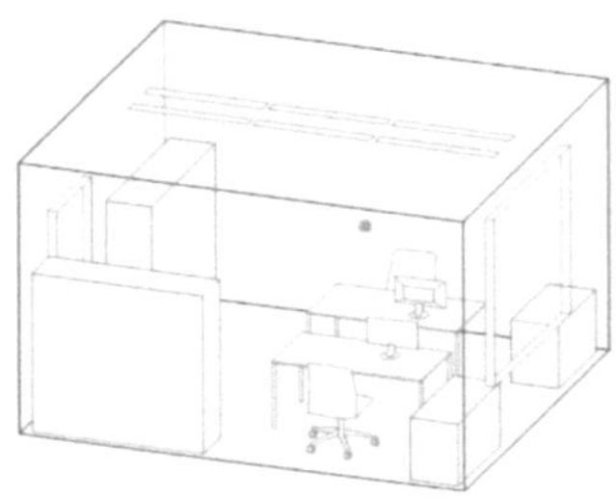

Figure 11.
Sensor positioning.

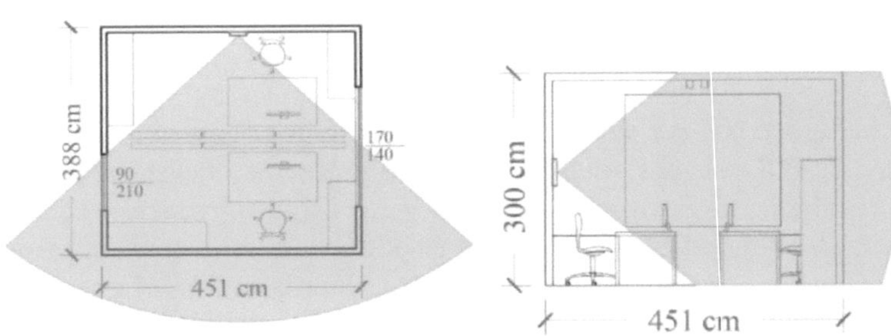

Figure 12.
Sensor radius of action.

The sensor was placed on the south wall and at a height that meets the datasheet requirement (**Figure 11**). The range of the sensor covers the area fairly accurately (**Figure 12**), although some areas remain uncovered. The main issue related to the sensor position concerns the presence detection. All other quantities are evaluated punctually at the sensor location.

4.4 Motion sensor FIBARO

The motion, light, and temperature sensor; motion sensor is a product of Fibaro, a Polish brand operating in the IoT (Internet of Things) sector and a home automation company. The sensor is small in size (**Figure 13**), easily installed, non-invasive in

DOI: http://dx.doi.org/10.5772/intechopen.114831

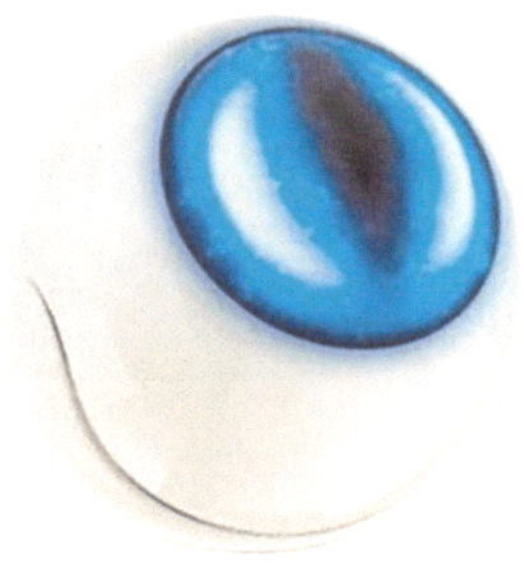

Figure 13.
Fibaro motion sensor.

	Quantity	Unit	Accuracy	Range	Sampling
	Temperature	[°C]	0.5°C (range 0–40°C)	−20 to 100°C	At least 1 report every 2 hour
	Presence	[on/off]	—	—	—
	Illuminance	[lx]	—	0 to 32,000 lx	At least 1 report every hour

Table 8.
Motion sensor specifications.

indoor environments, and wireless because battery-powered. In addition to detecting motion, temperature, and light intensity, the sensor has a built-in accelerometer that detects changes in position or evidence of device tampering. Detected motion, temperature, and vibration are reported through the LED embedded in the object.

The sensor, in terms of temperature, is within the range of required accuracy reported by EN ISO 7726 (**Table 8**) [2]. A height of 2.4 m is recommended for installation; however, the motion sensor was placed on the south wall at a height of 1.6 m near the rest of the sensors (**Figures 14** and **15**) to make an accurate comparison. Nevertheless, a lower position was chosen because a higher location would have distorted the temperature monitoring, which would not have been representative of the position of the office users.

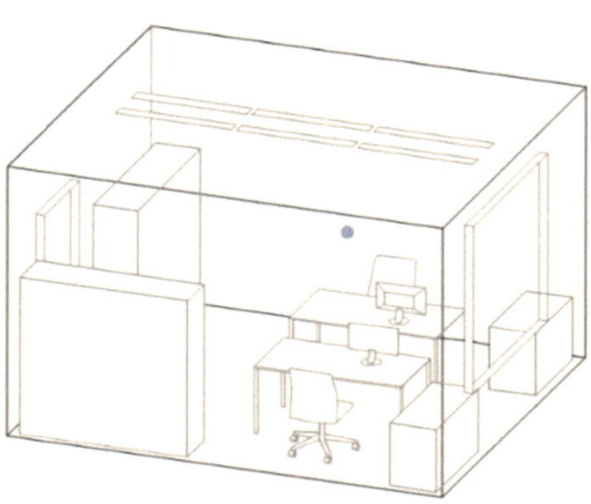

Figure 14.
Positioning of the sensor.

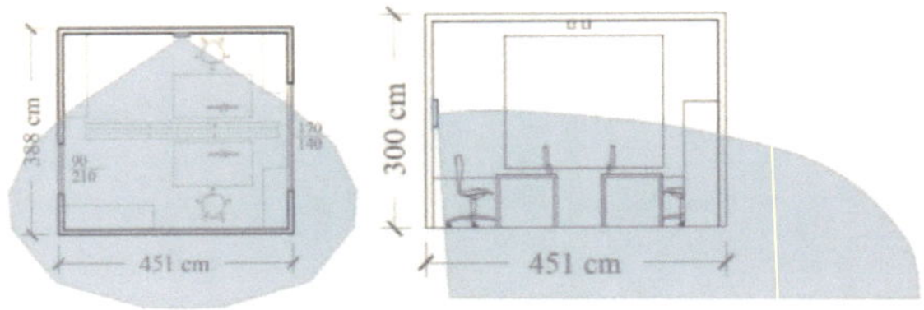

Figure 15.
Sensor radius of action.

4.5 Text 480: Air conditioning meter

Testo 480 is an instrument for measuring climatic parameters and is particularly suitable for thermal comfort analysis (**Figure 16**). Testo began as a small thermometer manufacturing company in 1957 and today is a company with a worldwide presence as a provider of measurement solutions. The instrument takes the form of a handheld with six connections for different probes, all placed on a stand. For each connected probe, the handheld displays a tab with the different quantities. The instrument has an autonomy of 17 hours thanks to the built-in battery, but it also has a power supply connection, data are saved in an internal memory and are downloadable *via* USB connection and text software to a personal computer.

The probes included for comfort assessments are:

- Globometer probe Ø 150 mm, used to measure radiant heat following ISO 72461, ISO 7726, DIN EN 277262, and DIN 334033. It has a type K thermocouple in the center to measure temperature and has an adaptation time of about 30 minutes.

- IAQ probe, integrated probe that measures CO2, humidity, temperature, and absolute pressure.

- Luxmetric probe, for measuring illuminance.

- Well-being probe, used to measure air velocity and determine the risk of drafts. It also allows for measuring the degree of turbulence, which corresponds to the magnitude of temporal fluctuations in air velocity, an important parameter for calculating the risk of drafts. This probe makes it possible to determine air velocity and draft risk as well as ambient temperature and pressure. It is a hot-wire technology probe, and with low air velocity, the displayed temperature is

Figure 16.
Testo 480.

slightly higher, justifying the reason for using the temperature values returned by the IAQ probe for analysis.

This climatic parameter meter was chosen as the comparison item because it is the one whose technical specifications fall within all that to be standardized (**Table 9**).

The probes were mounted on the provided tripod and placed in the center of the room. This choice was made to properly detect all the quantities by making them make sense within the panorama on comfort assessment. The height of the trestle was modulated about 1.3 m from the floor, a height similar to that of a seated person, and the center of the room was chosen basically for a correct measurement of the average radiant temperature. More correct measurements would have been made by placing an instrument near each workstation, so due to limitations related to instrumentation availability, an average position relative to the two workstations was chosen. The only probe that was placed differently from the others was the luxmeter. Regarding the measurement of illuminance, the difficulties encountered were different. It usually makes sense to calculate illuminance on a work surface and not on vertical walls. The problem arose because most of the sensors previously analyzed are integrated sensors, so the obvious and correct location for measuring one parameter is not necessarily correct for all the others. In light of this, the luxmeter of the control unit was placed on the south wall next to the other sensors, to at least have an illuminance value comparable with the others, and then choose the best position for the sensor in the demonstrators.

		Quantity	Unit	Accuracy	Sampling
	Temperature	[°C]	±0.5°C	0 – +50°C	30 report/h
	Relative Humidity	[%]	±(1.8%UR + 0.7% del v.m.)	0–100%	30 report/h
	Illuminance	[lx]	Classe C (DIN 5032-7[1])	0 – +100,000 lux	30 report/h
	CO_2	[ppm]	±(75 ppm + 3% del v.m.) 0 a + 5000 ppm ± (150 ppm + 5% del v.m.) 5001 a + 10,000 ppm	0 – +10,000 ppm	30 report/h
	Average radiant temperature	[°C]	Classe 1 (EN 60584-2[2])	0–120°C	30 report/h
	Air velocity	[m/s]	±(0.03 m/s + 4% del v.m.)	0 – +5 m/s	30 report/h
	Absolute pressure	[hPa]	±3.0 hPa	+700 – +1100 hPa	30 report/h

[1]*Norma nazionale tedesca. Fotometria – Parte 7: Classificazione dei misuratori di illuminamento e misuratori di luminanza.*
[2]*International Electrotechnical Commissione Termocoppie. Parte 2: Tolleranze. Questa pubblicazione è stata sostituita da IEC 60584-1: 2013.*

Table 9.
Testo 480 specifications.

4.6 CO2-display

The CO2-display instrument, which can be installed on walls or tables, is an interesting solution when the simultaneous measurement of CO_2, humidity, and temperature is required (**Figure 17**). Humidity measurement is done through the ROTRONIC HYGROMER® IN-1 sensor. The instrument is configurable *via* the side keyboard, and the recorded data can be downloaded to a USB stick and analyzed with free software. The object has dimensions 330 × 250 × 50 mm, DC 12 VDC power supply type, and an internal memory capable of recording up to 18,000 measured values (**Table 10**).

The sensor was placed on the southernmost desk near the Testo system (**Figure 18**). This tool, too, should be an accurate control and comparison system for testing different commercial sensors that could be used in different indoor applications.

Figure 17.
CO2-display.

	Quantity	Unit	Accuracy	Range	Sampling
	Temperature	[°C]	±0.3 °C	−20 – +50°C	30 report/h
	Relative Humidity	[%]	±0.3% (a 25°)	0–100%	30 report/h
	CO_2	[ppm]	±30 ppm ±5% del valore misurato	0–5000 ppm	30 report/h

Table 10.
Technical characteristics CO2-display.

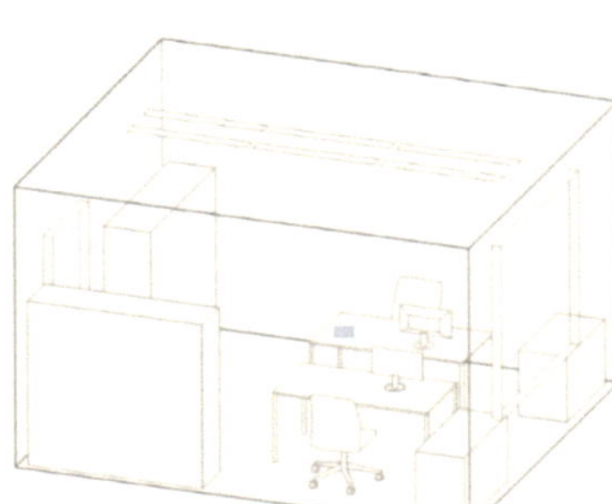

Figure 18.
Positioning of the sensor.

5. Analysis of winter case

After having fine-tuned the methodology, the data acquired from all the sensors installed during the winter period were compared to be able to choose the most suitable for field applications. The Sesto Senso sensor took over, and ERS-CO_2 replaced Apio-Sense. Assuming as reference the Testo 480 control unit, the data returned by Sesto Senso, in addition to having a completely similar trend, almost showed complete overlap. Even the ERS CO_2 sensor although it has a phase shift of half a degree, can be considered sufficiently accurate. The data that arouse much suspicion are those for the motion sensor from Fibaro, which returned more sparse data. Later, it was verified that the problem was not strictly with the sensor but related to the network that handled its communication with the data storage platform.

From the 12 days of data collected, **Figure 19** shows detail on the monitored days, precisely Thursday, December 5, 2019 to Saturday, December 7, 2019. We can see that with almost all sensors we fall within that temperature range dictated by Italian regulations, in particular the Presidential Decree April 16, 2013 No. 74 (19–22°C). The peak in temperatures represents the hours when the room is occupied. This probably occurs due to the sum of rising outdoor temperatures, radiative effects, and indoor thermal inputs.

As for relative humidity (**Figure 20**), it can be appreciated that in the hours of occupancy, there is a lowering of the monitored values. Thus, despite the internal

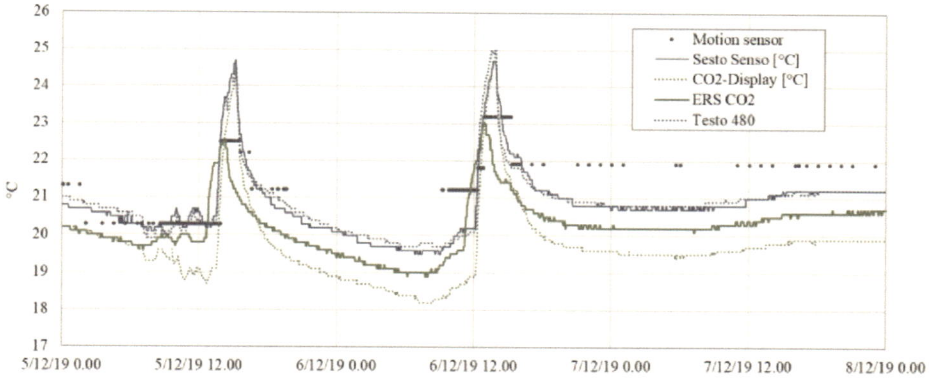

Figure 19.
Temperature recorded from 5/12/2019 to 8/12/2019.

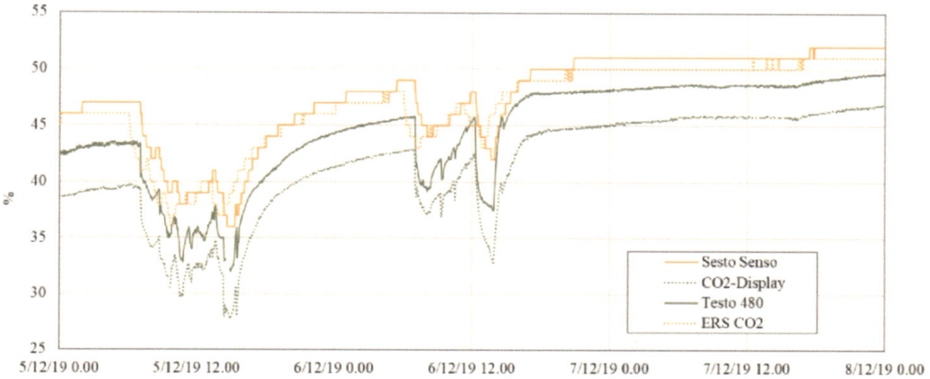

Figure 20.
Relative humidity recorded from 5/12/2019 to 8/12/2019.

loads, the rise in temperatures leads to a parameter reduction. We can also see that Sesto Senso and the ERS CO2 sensor, despite an overestimation of less than 5% in any case, are the ones that come closest to the reference values.

Another key parameter for assessing the environment is CO_2 concentration. Three sensors were available for its assessment (**Figure 21**). The Testo 480 instrument and the CO_2-display agree with each other for the readings, although with a slight discrepancy. Sixth Sense, in this case, is extremely poor in its ratings. The underlying causes of the malfunction are to be found in the relief of all those quantities that contribute to the calculation of CO2.

The first problem encountered is in the evaluation of people in the room. The sensor installed at the door, as presented in the paragraph, in addition to counting the actual entry and exit of people, also includes opening and closing the door. Another problem that distorted the evaluation was the detachment of the window opening and closing sensor. Therefore, the system, and probably the algorithm, will need an intervention.

In contrast, the most difficult quantity to detect and compare was the illuminance. The first difficulty was encountered in deciding where to place the sensor. All sensors return illuminance in lux; this quantity must be used to assess whether there is the right degree of illuminance on a work surface. Since almost all the sensors used are integrated, it was necessary to find the right compromise for surveying the different magnitudes. In addition, after an initial analysis, it was seen that the sensors returned conflicting data, so it was first thought to assess the data.

Figure 22 depicts that curves are quite different from each other, moreover, the recorded illuminance values are also significantly lower than the recommended ones (between 200 and 350 lux). Considering the Testo instrument as a reference, a deeper analysis was done to understand their trend.

From the Sesto Senso sensor, it is worth noting to detect a trend line and a possible corrective factor (**Figure 23**), but for the other two sensors, more in-depth calibrations would be needed.

Mean radiant temperature and air velocity were evaluated only with the Testo 480 control unit aimed at parametric analysis on the PMV calculation and identify possible simplifications of the calculation in cases where these quantities are not available. Regarding the air velocity (**Figure 24**), the range accepted by the standard is highlighted in green, and it can be appreciated that the greatest fluctuations are in the

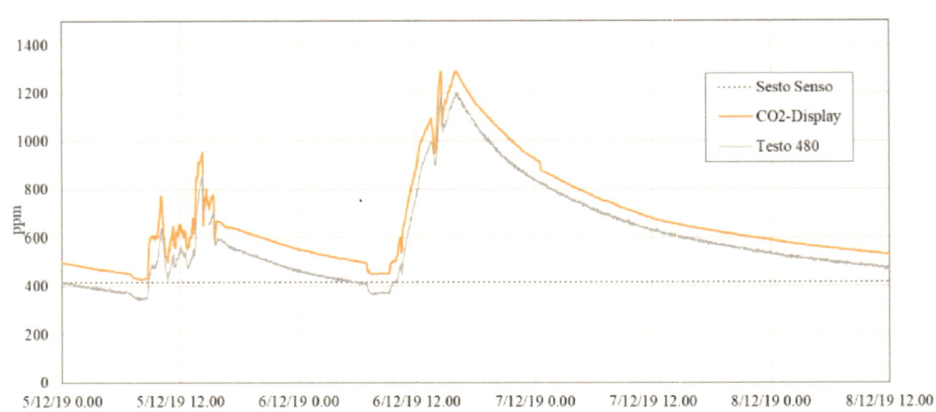

Figure 21.
CO2 recorded from 5/12/2019 to 8/12/2019.

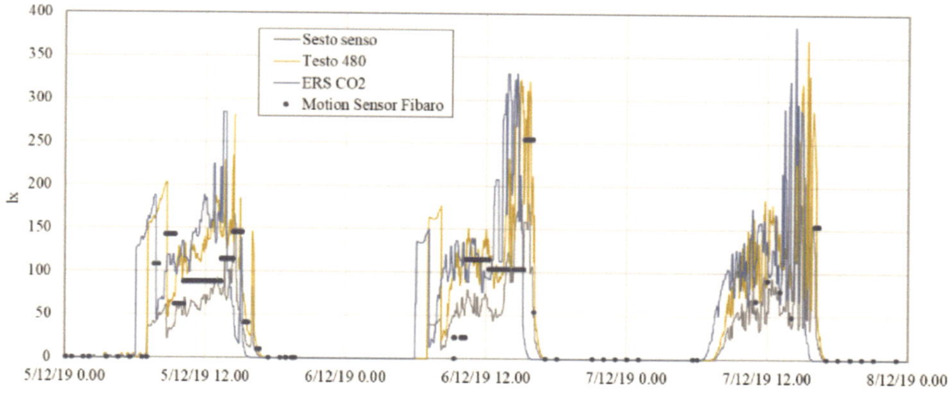

Figure 22.
Illuminance recorded from 5/12/2019 to 8/12/2019.

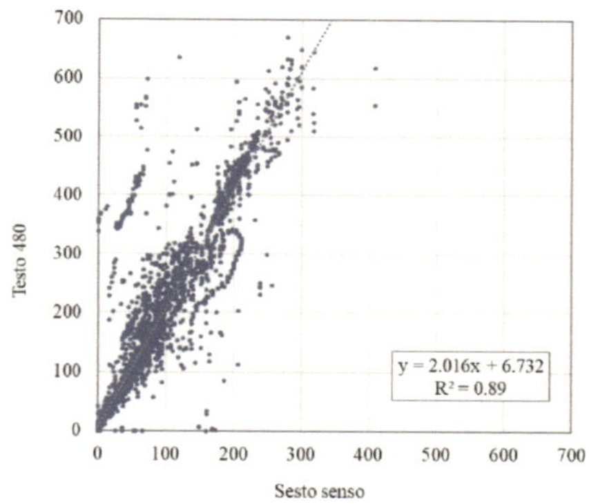

Figure 23.
Sensors correlation.

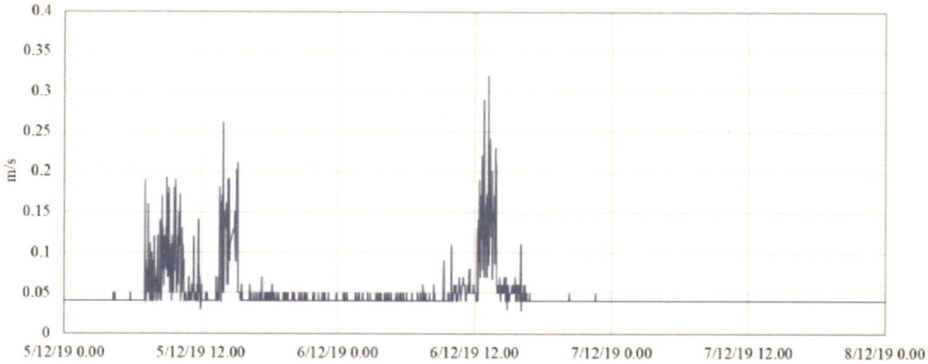

Figure 24.
Wind data recorded from 5/12/2019 to 8/12/2019.

hours of occupancy, perhaps due to both the opening of doors and windows and the passage and movement of people. The average radiant temperature (**Figure 25**), on the other hand, undergoes significant increases in the presence of occupants.

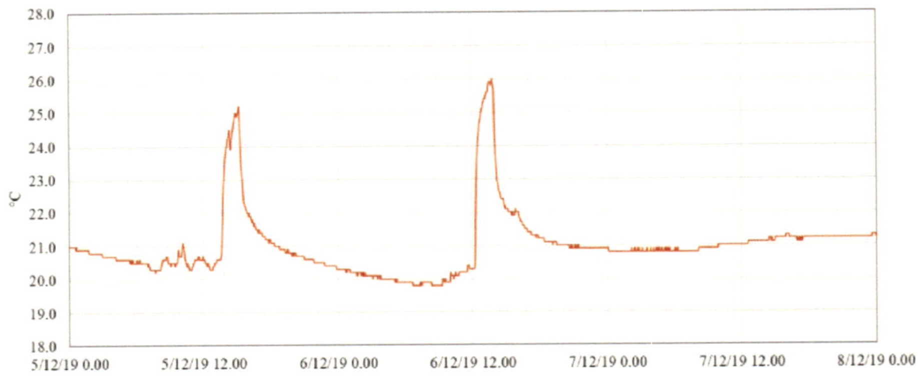

Figure 25.
Mean radiant temperature data recorded from 5/12/2019 to 8/12/2019.

6. Evaluation of comfort indexes

The comfort indices evaluated refer to winter data only, because of the greater accuracy in the measurements and because there would be no substantial difference from the results obtained in summer. From the analysis of the sensors and the magnitudes they return, three possible scenarios were identified (**Figure 26**).

In the summer case, the Humidex [10] was considered in addition to the conventional PMV index (**Figure 27**). The Humidex Index is the bioclimatic index initially used in Canada that attempts to describe the physiological discomfort caused by the combination of heat and humidity in the air. The numerical value represents an apparent temperature that provides indications of the severity of a climatic condition. It should be used only for temperature ranges between 21°C and 55°C, and reference is made to the scale in **Table 11**.

This index is based on a simple empirical relationship that takes into account air temperature and vapor pressure. Actually, instead of relative humidity, this index uses a related parameter, namely vapor pressure, according to the formula:

$$H = T_a + (0.5555\ (e - 10)) \tag{8}$$

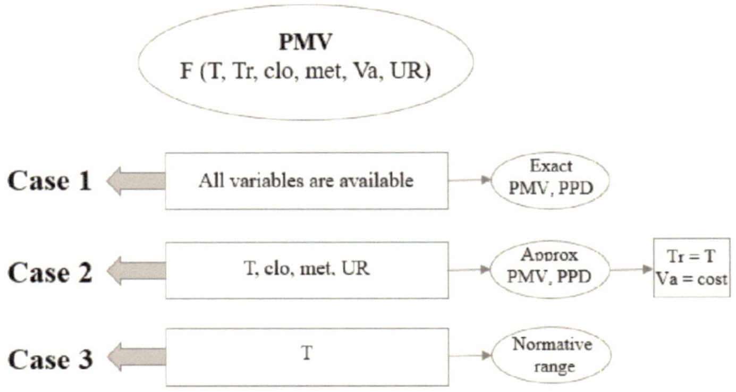

Figure 26.
Possible scenarios from comfort indexes evaluation.

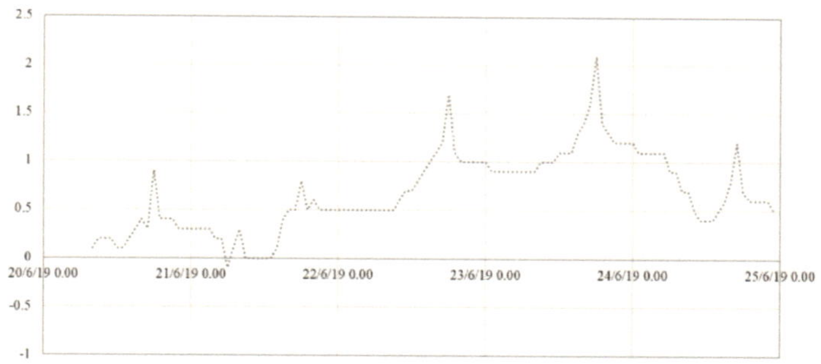

Figure 27.
PMV evaluated with testo 480.

To 27°C	No discomfort
From 27 to 30°C	Feeling of Discomfort
From 30 to 40 °C	Intense discomfort. Caution: limit heavier physical activities
From 40 to 55°C	Severe feeling of discomfort. Danger: avoid efforts
Over 55 °C	Danger of death: imminent heat stroke

Table 11.
Reference scale for humidex index [10].

where precisely, Ta = temperature in °C, while e = air vapor pressure (hPa). Since it is easier to know relative humidity (RH) rather than vapor pressure, by knowing the relation between these parameters, another version of the Humidex index can be formulated:

$$H = T_a + \left(0.5555\left(0.06^{*}\ UR^{*}\ 10^{0.03Ta} - 10\right)\right) \quad (9)$$

In **Figure 28** can be seen that with the Humidex index, there is a slight overestimation of perceived discomfort.

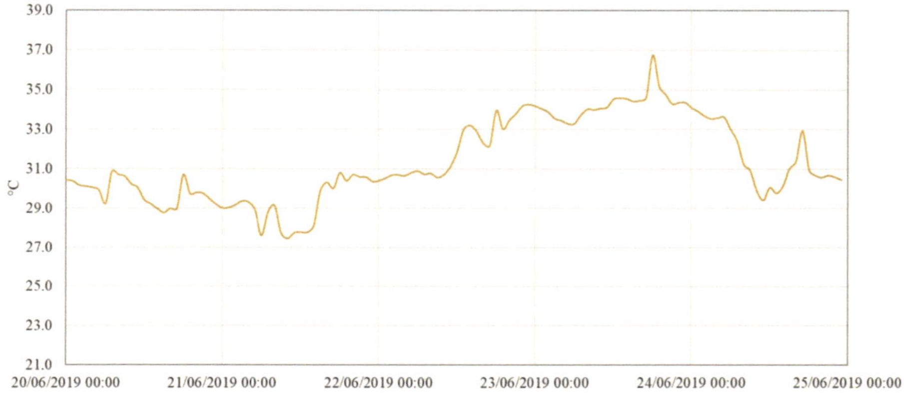

Figure 28.
Humidex evaluated with testo 480.

INPUT							OUTPUT					
Abbigliamento (clo)	Temp. dell'aria(°C)	Temp. Media radiante (°C)	Metabolismo energetico (met)	Velocità dell'aria (m/s)	Umidità relativa (%)	Lavoro esterno (met)	Temp. operativa (°C)	PMV	PPD	Iterazioni	Data	Ora
1,00	23,5	24,9	1,2	0,05	53,7	0	24,2	0,6490	13,8438	6	28/11/2019	15:48
1,00	23,5	24,9	1,2	0,05	55,0	0	24,2	0,6573	14,0760	6	28/11/2019	15:50
1,00	23,6	24,9	1,2	0,04	54,3	0	24,25	0,6658	14,3137	6	28/11/2019	15:52
1,00	23,5	24,9	1,2	0,05	52,5	0	24,2	0,6412	13,6322	6	28/11/2019	15:54
1,00	23,6	24,8	1,2	0,04	52,4	0	24,2	0,6436	13,6965	6	28/11/2019	15:56
1,00	23,5	24,8	1,2	0,04	51,9	0	24,15	0,6275	13,2622	6	28/11/2019	15:58
1,00	23,6	24,7	1,2	0,04	51,7	0	24,15	0,6291	13,3070	6	28/11/2019	16:00
1,00	23,6	24,7	1,2	0,05	51,4	0	24,15	0,6272	13,2552	6	28/11/2019	16:02
1,00	23,6	24,7	1,2	0,07	51,4	0	24,15	0,6272	13,2552	6	28/11/2019	16:04
1,00	23,7	24,7	1,2	0,04	51,0	0	24,2	0,6374	13,5291	6	28/11/2019	16:06
1,00	23,8	24,8	1,2	0,05	50,5	0	24,3	0,6569	14,0625	6	28/11/2019	16:08
1,00	23,8	24,8	1,2	0,06	51,0	0	24,3	0,6601	14,1542	6	28/11/2019	16:10
1,00	23,9	24,8	1,2	0,05	50,2	0	24,35	0,6676	14,3656	6	28/11/2019	16:12
1,00	23,9	24,8	1,2	0,04	49,8	0	24,35	0,6650	14,2909	6	28/11/2019	16:14
1,00	24,0	24,8	1,2	0,04	49,9	0	24,4	0,6784	14,6726	6	28/11/2019	16:16
1,00	24,0	24,8	1,2	0,04	50,3	0	24,4	0,6810	14,7493	6	28/11/2019	16:18
1,00	24,0	24,8	1,2	0,07	49,6	0	24,4	0,6764	14,6153	6	28/11/2019	16:20

CALCOLA

Range accettabile	
Clothing	[0 to 2clo]
Air temp	[10 to 30°C]
Mean radiat temp	[10 to 40°C]
Activity	[0.8 to 4met]
Air speed	[0 to 1m/s]
Relative Humidity	[30 to 70%]
Lavoro esterno	[0 to 0,2met]

Figure 29.
Excel spreadsheet for calculation of PMV and PPD.

As for the PMV index, it was calculated using an Excel spreadsheet (**Figure 29**) in which with the help of a proper macro, the program in basic provided by EN ISO 7730 [3] was implemented.

The first identified case involves the exact calculation of PMV, which can only be obtained if an instrument, such as the Testo 480 control unit, that returns all the necessary quantities is available. In (**Figure 30**) the trend of PMV for occupancy hours only and the percentage of dissatisfaction is depicted. From the processing of these data, it can be appreciated that for 74% values fall within $-0.5 < \text{PMV} < 0.5$ and for 25% this range is exceeded, but remains between -1 and $+1$, still acceptable. As for the percentage of dissatisfied, on the other hand, it touches 15%.

The second case involves an approximation of those variables that are more complex to calculate, and that commercial home automation sensors rarely measure, such as the mean radiant temperature and the air velocity. In rooms that do not have large glass surfaces, the difference between air temperature and mean radiant temperature is very small [11, 12]. **Figure 31** shows that the difference between the two quantities rarely exceeds half a degree, and the mean radiant temperature disagrees by less than 3% with the air temperature (see **Figure 32**). It was thus thought to replace the average radiate temperature with the air temperature. Another quantity that is hardly measured in home or work environments is air velocity. Both mean radiant temperature and air velocity would also have to be measured at several points in the room, near each workstation, which is quite complex. **Figure 33** shows that during the hours of occupancy, only in 0.07% of cases the air velocity exceeds 0.1 m/s, with an average of 0.05 m/s. In light of these considerations, the PMV slightly varies by replacing the

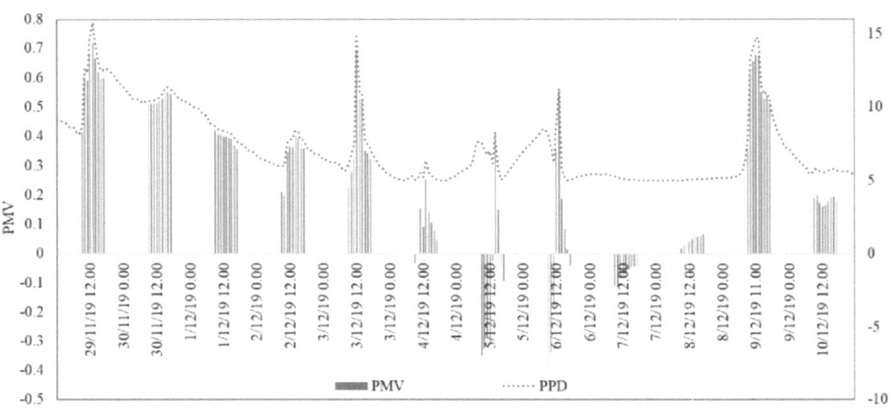

Figure 30.
PMV e PPD evaluated with testo 480.

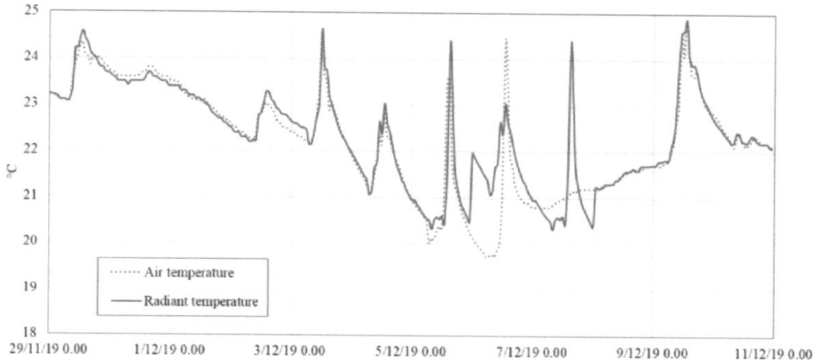

Figure 31.
Mean radiant temperature and air temperature comparison.

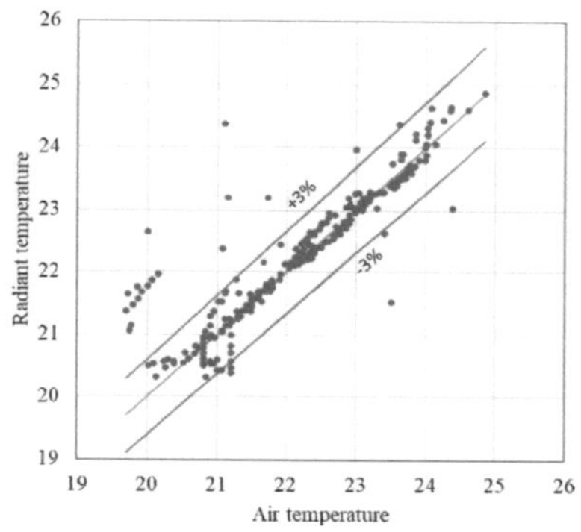

Figure 32.
Evaluation of average radiant temperature/air temperature error.

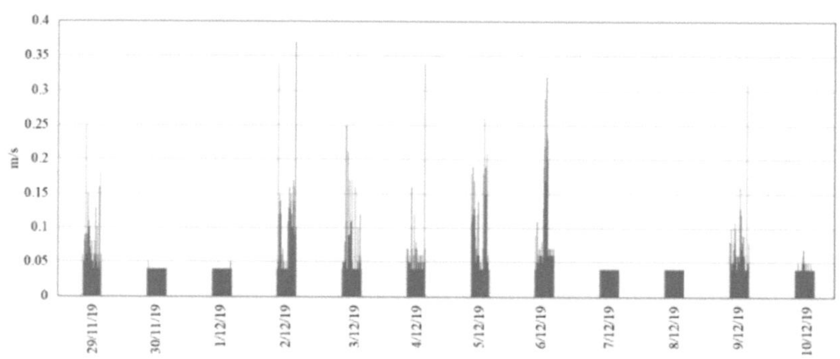

Figure 33.
Air velocity during periods of occupancy.

average radiant temperature with the air temperature and placing the air velocity equal to 0.05 m/s (**Figure 34**).

Despite the approximations, (**Figure 35**) highlights that the error is minimal and R^2 is almost 1, so in rooms that do not have radiant floors/ceilings/walls and that do not have large windows, the reduction of variables is possible.

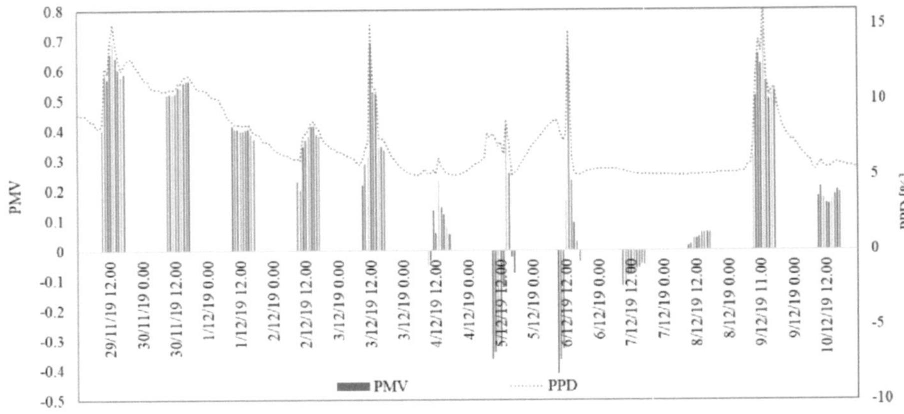

Figure 34.
Approximate calculation of PMV and PPD.

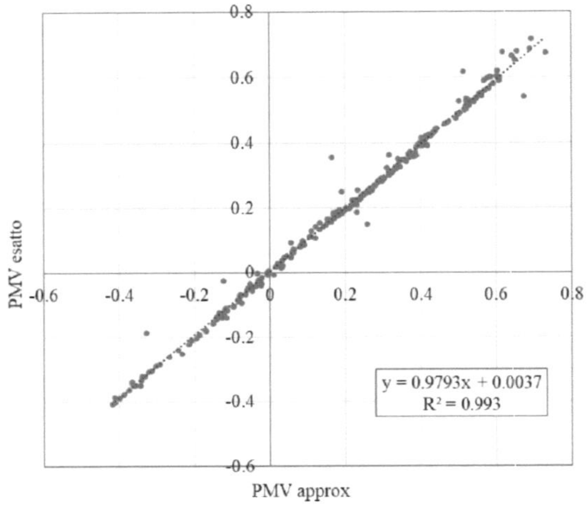

Figure 35.
Approximated vs. exact PMV.

7. Conclusions

Contemporary urban society often confines individuals to indoor spaces, whether residential or occupational, for significant portions of their lives. In recent years, there has been a growing interest in improving indoor environmental quality (IEQ), driven by heightened awareness of its impact on human health and productivity. This study leverages experimental data analysis to assess and identify key variables influencing comfort perception, aiming to streamline calculation methodologies. Special attention is directed toward the "Sesto Senso" sensor, for which a prototype was available. Evaluation of the sensor's measurements revealed discrepancies, particularly in the algorithm calculating CO2 levels, traced back to issues with presence sensors and door/window opening/closing sensors. Repositioning and recalibration of these sensors were performed to address the discrepancies. Additionally, an upgrade to the Sesto Senso system was implemented, incorporating a predicted mean vote calculation for thermo-hygrometric comfort assessment, currently undergoing testing at the Casaccia center.

Author details

Daniela Cirone[1]*, Sabrina Romano[2], Roberto Bruno[1] and Natale Arcuri[1]

1 University of Calabria, Rende, Italy

2 ENEA Italian National Agency for New Technologies, Energy and Sustainable Economic Development, Rome, Italy

*Address all correspondence to: daniela.cirone@unical.it

References

[1] Fan L, Han X, Li L, Liu H, Ge T, Wang X, et al. Indoor air quality of urban public transportation stations in China: Based on air quality evaluation indexes. Journal of Environmental Management. 2024;**349**:119440

[2] EN ISO 7726: 2012; Ergonomics of the Thermal Environment. Instruments for measuring physical quantities. Technical Committee: ISO/TC 159/SC 5, ICS: 13.180

[3] UNI EN ISO 7730:2005; Ergonomics of the thermal environment. Analytical determination and interpretation of thermal comfort using calculation of the PMV and PPD indices and local thermal comfort criteria.Technical Committee: CEN/TC 122 - Ergonomics, Drafting Committee: CEN/TC 122/WG 11 - Ergonomics of the Physical Environment, ICS: 13.180

[4] ISO 7933:2023; Ergonomics of the thermal environment. Analytical determination and interpretation of heat stress using calculation of the predicted heat strain. Technical Committee: ISO/TC 159/SC 5. ICS: 13.180

[5] ISO 11079:2007; Ergonomics of the thermal environment. Determination and interpretation of cold stress when using required clothing insulation (IREQ) and local cooling effects. Technical Committee: ISO/TC 159/SC 5, ICS: 13.180

[6] Atmaca I, Kaynakli O, Yigit A. Effects of radiant temperature on thermal comfort. Building and Environment. 2007;**42**(9):3210-3220

[7] Zhang H, Yang R, You S, Zheng W, Zheng X, Ye T. The CPMV index for evaluating indoor thermal comfort in buildings with solar radiation. Building and Environment. 2018;**134**:1-9

[8] Cannistraro M, Trancossi M. Enhancement of indoor comfort in the presence of large glazed radiant surfaces by a local heat pump system based on Peltier cells. Thermal Science and Engineering Progress. 2019;**14**:100388

[9] Djongyang N, Tchinda R, Njomo D. Thermal comfort: A review paper. Renewable and Sustainable Energy Reviews. 2010;**14**(9):2626-2640

[10] Masterton JM, Richardson FA. Humidex: a method of quantifying human discomfort due to excessive heat and humidity. Report, Environment Canada, No. No. CLI 1-79, 45 pp ref. 13

[11] Wang D, Chen G, Song C, Liu Y, He W, Zeng T, et al. Experimental study on coupling effect of indoor air temperature and radiant temperature on human thermal comfort in non-uniform thermal environment. Building and Environment. 2019;**165**:100388

[12] Zhou X, Liu Y, Luo M, Zhang L, Zhang Q, Zhang X. Thermal comfort under radiant asymmetries of floor cooling system in 2 h and 8 h exposure durations. Energy and Buildings. 2019; **188**:98-110

Chapter 2

Indoor Risk Estimation Model for Health Protection from COVID-19

Sandra Costanzo and Alexandra Flores

Abstract

In this chapter, a conceptual framework for assessing the probability of COVID-19 transmission is presented and discussed. The Wells-Riley probabilistic methodology for indoor environments is adopted, by integrating updated clinical data, thereby ensuring a reliable estimation of the probability of infection, through the model implementation on a specific Android platform. Notifications are sent to the user when detecting a high probability of infection and high carbon dioxide concentration levels. In addition, the Bluetooth signal is exploited to accurately determine the proximity between devices, thus facilitating the efficient enforcement of social distancing protocols among individuals. The effectiveness of the proposed model is validated through the application on different test scenarios.

Keywords: COVID-19, smart healthcare, indoor risk estimation, Wells-Riley model, distance measurement, RSSI, bluetooth

1. Introduction

The Coronavirus disease (COVID-19) pandemic has highlighted the critical need for comprehensive public health protection measures, particularly when considering indoor environments with a high risk of viral transmission. The cause of COVID-19 is attributed to the severe acute respiratory syndrome coronavirus 2 (SARS-CoV-2) virus [1]. The transmission of the virus occurs through the dissemination of tiny liquid droplets from the oral or nasal cavities of an infected individual. These droplets are typically released during coughing, sneezing, speaking, singing, or breathing. These particles exhibit a spectrum of sizes, ranging from relatively sizable respiratory droplets to diminutive aerosols [2]. The SARS-CoV-2 viral load in the oral cavity has been observed to exhibit a wide range of values, as reported in Refs. [3, 4], lasting from 10^2 to $10^1$1 copies per mL of respiratory fluid. The viral loads exhibit fluctuations throughout the illness progression, with a tendency to reach their highest levels near the symptoms manifestation [5]. Factors such as indoor gatherings and inadequate ventilation may facilitate the spread of COVID-19. A comprehensive analysis of over 300 COVID-19 outbreaks, each one involving three or more individuals, has revealed that all instances of transmission are associated with specific indoor settings [6]. According to a research study adopting contact tracing [7], the probability of COVID-19 transmission from a primary case is 18.7 times higher in an enclosed environment as compared to an outdoor environment. In order to estimate the potential for COVID-19 transmission within indoor areas and to execute practical mitigation efforts, it is crucial to develop reliable risk estimation models. These should be founded on theories and mathematical equations that are biologically plausible or consistent with clinical or laboratory evidence. In general, assessing the risk of infection involves two primary components: first, estimating the amount of the infectious agent that has been ingested, and second, estimating the probability of infection based on this amount. Deterministic and stochastic models can be used to classify infection risk. Each individual has a certain tolerance level for the infectious agent, and if they are exposed to a dose that exceeds their tolerance level, they will become infected. By this hypothesis, the deterministic models can predict whether or not an individual will be infected after a given dose of ingestion. Instead, stochastic models estimate the probability of contracting the infection at the ingested dose [8]. The Wells-Riley model is extensively applied to estimate the risk of airborne infectious disease transmission. It provides a framework for computing the risk of infection based on essential variables in a particular environment. Recently, it has been widely utilized to determine the COVID-19 transmission rate [9]. This model, originally developed by William F. Wells and Richard L. Riley [10, 11], was initially used to investigate tuberculosis and measles. In 1995, Well expressed the viral load emitted in terms of the quantum emission rate ($E, quanta/h$). Quantum refers to the number of infectious airborne particles needed to infect a person. On the other hand, in 1978, Riley proposed a modification of the Reed-Frost equation to estimate the probability of averting infection based on the amount (in quanta) of airborne pathogen particles a person intakes. In order to mitigate the transmission of COVID-19, it is essential to consider some factors [12], including social distancing, interpersonal contact, mask adherence, and ventilation. Adhering to the recommended guidelines of health authorities to maintain a safe physical distance from others is paramount in mitigating transmission risks. Reducing social gatherings and/or adopting the preference for virtual alternatives, may mitigate potential exposure. Properly donning masks to cover both the nose and mouth can offer an extra level of safeguarding. Adequate ventilation, especially in enclosed environments, may facilitate the dispersion and the elimination of suspended particulate matter in the air.

The initial focus of this chapter relates to the theoretical aspect of the Wells-Riley model, which is adapted to determine the risk of infection by COVID-19. The methodology to acquire each employed parameter is also discussed. Section 3 includes the modeling and the analysis of the outcomes derived from the model. Section 4 exposes the architecture of an experimental setup, which aims to estimate the probability of infection and the degree of distancing between individuals within indoor environments. Section 5 displays the outcomes of the experimental configuration for different scenarios. Finally, in Section 5, the conclusions are outlined.

2. Theory and formulation

The Wells-Riley [10, 11] equation is formulated by combining the Poisson probability distribution, where the probability of infection p is dependent on the total inhaled infectious quanta n. Inhaled quanta are determined by multiplying the quanta concentration of the air $\left(C_{avg}, quanta/m^3\right)$ by the pulmonary ventilation volumetric flow rate of the exposed person $\left(Q_b, m^3/h\right)$ and the duration of exposure in hours (D, h), as follows:

$$p = 1 - e^{-n} = 1 - e^{-C_{avg} \cdot Q_b \cdot D} \quad (1)$$

Assuming equilibrium and well-mixed conditions, Eq. (2) describes the quantum concentration C in a room with a ventilation rate Q, housing I infected individuals who emit infectious pathogens at a constant rate of q infectious quanta per person per unit time.

$$C_{avg} = \frac{l \cdot q}{Q} \quad (2)$$

Therefore, the probability of infection for a person in this room is calculated with Eq. (3), which can be also represented as the relationship between the number of cases of developing infection N_{ca} and the number of susceptible people N_{sus}.

$$p = 1 - e^{\frac{-l \cdot q \cdot Q_b \cdot D}{Q}} = \frac{N_{ca}}{N_{sus}} \quad (3)$$

2.1 Quanta emission rates for SARS-COV-2

The emission rate of quanta is highly variable, ranging from 3 to 300 quanta per hour, as reported in [13], whereas in [14], based on the work of [15], it is recommended a range of ~ 2–14 quanta per hour for oral breathing and 61–408 quanta per hour for loud speaking. This variation is influenced by the intensity of activities, such as shouting, chanting, speaking aloud, and metabolic rates. The values relative to four different activity levels (resting, light intensity, moderate intensity, and high intensity) are shown in **Table 1**, based on reference [14]. The quanta emission rate is calculated by assuming the droplets of an infected person contain the same viral load as the sputum [15]. Using a mass balance, the expelled viral load can be determined from the virus concentration in the sputum and the number of tiny droplets smaller than 10 μm in size. The emitted viral load is expressed as the quantum emission rate (ERq, quanta h1) and evaluated as follows:

$$ER_q = c_v \cdot c_i \cdot V_{br} \cdot N_{br} \cdot \int_0^{10\mu m} N_d(D) \cdot dV_d(D) \quad (4)$$

where c_v is the viral load in the sputum (measured as RNA copies per mL^{-1}), c_i is a conversion factor that relates an infectious quantum to the infectious dose expressed in viral RNA copies, V_{br} is the volume of exhaled air per breath measured in cm^3, N_{br} is the breathing rate in breaths per hour breath h^{-1}, N_d is the droplet number

Activity	Oral breathing	Speaking	Aloud speaking or singing
Sedentary/passive resting	2	9.4	60.5
Light intensity/standing	2.3	11.4	65.1
Moderate intensity	5.6	26.3	170
High intensity	13.5	63.1	408

Table 1.
Quanta emission rates for SARS-CoV-2 (quanta/hour) [14].

concentration in (part. cm^3) and V_d is the volume of a single droplet in mL, which is determined by the diameter D of the droplet. The volume of the drop V_d is calculated using experimental data obtained by the authors of [16].

2.2 Volumetric breathing rates

Estimating the volume of air exchanged during breathing is crucial for understanding the factors that contribute to the spread of respiratory droplets and aerosols. Airborne respiratory droplets containing germs are released whenever an infected person breathes, coughs, or speaks. This study employs the Exposure Factors Manual (Chapter 6 of the United States Environmental Protection Agency [14]) factors. **Table 2** displays information on average daily inhalation rates for short exposures as a function of age and activity level.

2.3 Mask efficiencies in reducing virus emission

The investigation also focuses on the adoption of respiratory protective equipment, such as masks, by both infected and susceptible individuals. Masks provide a physical barrier that can capture respiratory droplets harboring viruses, restricting their escape into the environment. The effectiveness of various masks in reducing virus emission is compared in **Table 3**.

2.4 Ventilation rates

The estimation of the infection rate caused by aerosols in an indoor environment depends on a crucial parameter, namely the ventilation rate. This uniquely refers to the substitution of indoor air with outdoor air. As indicated in Ref. [14], a ventilation rate of one hour does not necessarily imply a complete air replacement within the

Activity level	Age group	Mean	95*th* Percentile
	[years]	$m^3/minute$	$m^3/minute$
Sedentary/passive	16 to <21	5.3E–03	7.2E–03
	21 to <31	4.2E–03	6.5E–03
	31 to <41	4.3E–03	6.6E–03
Light intensity	16 to <21	1.2E–02	1.6E–02
	21 to <31	1.2E–02	1.6E–02
	31 to <41	1.2E–02	1.6E–02
Moderate intensity	16 to <21	2.6E–02	3.7E–02
	21 to <31	2.6E–02	3.8E–02
	31 to <41	2.7E–02	3.7E–02
High intensity	16 to <21	4.9E–02	7.3E–02
	21 to <31	5.0E–02	7.6E–02
	31 to <41	4.9E–02	7.2E–02

Table 2.
Short-term exposure values average daily volumetric breathing rates [14].

Mask	Exhalation mask efficiency	Inhalation mask efficiency
Type	**Infected person**	**Susceptible person**
N95 masks (KN95, FF2)	90%	90%
N95 with exhalation valve	0%	0%
Cloth, surgical	50%	30%
Face shields worn without a mask	23%	23%

Table 3.
Mask efficiency in preventing virus inhalation by a susceptible individual [14].

	Default values							
	People outdoor		Area outdoor		Occupant	Combined outdoor air		
	Air rate (Rp)		Air rate (Ra)		Density	Rate		
Occupancy	**cmf/**	**L/s.**	**cmf/ft^2**	**L/s.m^2**	**#/1000 ft^2**	**cmf/**	**L/s.**	**Air**
Category	**Person**	**Person**			**or #/100 m^2**	**Person**	**Person**	**Class**
Domestic	5	2.5	0.06	0.3	15	9	4.5	1
Schools	10	5	0.12	0.6	35	13	6.7	1
Food service	7.5	3.8	0.18	0.9	70	10	5.1	2
Hotels resorts/dormitories	5	2.5	0.06	0.3	10	11	5.5	1
Office buildings	5	2.5	0.06	0.3	10	11	5.5	1
Public assembly spaces	5	2.5	0.06	0.3	150	5	2.7	1
Sports and Entertainment	20	10	0.18	0.9	7	45	23	2

Table 4.
Minimum ventilation rates in breathing zone.

same period. The mixing process prevents the displacement of indoor air by fresh air. Ventilation variations across diverse settings are of significant importance.

ASHRAE Standard 62 [17] outlines the minimum ventilation rates required for conditioned spaces to ensure satisfactory indoor air quality (IAQ), taking into account the use and occupancy of the building. These rates are detailed in **Table 4**.

2.5 Decay rate of virus infectivity in aerosol

Environmental circumstances may influence the infectiousness of SARS-CoV-2 in natural aerosols, and the virus may be able to survive for extended periods in some conditions. Important factors that affect the persistence of infectious SARS-CoV-2 in aerosols include temperature, sunlight, and humidity. On the other hand, sunshine and temperature have a much more significant impact on the decomposition rate than humidity. According to research in [18], the time needed to reduce infectious virus by 90% ranged from 4.8 minutes at 40°C, 20% relative humidity, and high-intensity sunlight. In contrast, the time required indoors or at night is over two hours. The decay rate of aerosol virus infectivity is computed using the following equation:

$$K_{infectivity} = 0.16030 + 0.04018\left(\frac{T - 20.615}{10.585}\right) + 0.02176\left(\frac{RH - 45.235}{28.665}\right) + 0.14369\left(\frac{S - 0.95}{0.95}\right) + 0.02636\left(\frac{T - 20.615}{10.585}\right)\left(\frac{S - 0.95}{0.95}\right) \quad (5)$$

where $K_{infectivity}$ is the decay constant for viral infectivity; T is the temperature measured in degrees Celsius (°C); RH is the relative humidity (%) and S is the UV irradiance (W/m^2), ranging from 0 (indoors) up to 10 (full sun noon).

2.6 Virus removal rate using controlled systems

A useful indicator of ventilation quality is the rate of air change in the room. This measurement is known as air changes per hour (ACH). Ventilation with clean outdoor air removes viruses, particulates, and gasses, making it healthier. However, most heating and ventilation systems use recirculated air, which does not reduce the risk of COVID-19 unless the recirculated air is filtered to remove minute particles. An air filter's Minimum Efficiency Reporting Values (MERV) rating indicates how effectively it extracts airborne particles. MERV is expressed on a 16-point scale based on three ranges of Average Particle Size Efficiency (PSE) [19]. Refer to **Table 5** for MERV Parameters. For instance, if the filter has a MERV value of 12 or higher, it

Standard 52.2	Composite average particle size efficiency, size range, (μ m)			
Minimum	Range 1	Range 2	Range 3	Average arrestance,
Efficiency value	(0.3–1.0)	(1.0–3.0)	(3.0–10.0)	(%)
(MERV)				
1	n/a	n/a	E3 < 20	Aavg <65
2	n/a	n/a	E3 <20	65≤Aavg < 70
3	n/a	n/a	E3 <20	70≤Aavg <75
4	n/a	n/a	E3 <20	75≤Aavg
5	n/a	n/a	20 ≤E3	n/a
6	n/a	n/a	35 ≤E3	n/a
7	n/a	n/a	50 ≤E3	n/a
8	n/a	20 ≤E2	70 ≤E3	n/a
9	n/a	35 ≤E2	75 ≤E3	n/a
10	n/a	50 ≤E2	80 ≤E3	n/a
11	20 ≤E1	65 ≤E2	85 ≤E3	n/a
12	35 ≤E1	80≤E2	90 ≤E3	n/a
13	50 ≤E1	85≤E2	90 ≤E3	n/a
14	75 ≤E1	90≤E2	95 ≤E3	n/a
15	85 ≤E1	90≤E2	95 ≤E3	n/a
16	95 ≤E1	95≤E2	95 ≤E3	n/a

Table 5.
MERV (Minimum Efficiency Reporting Value) parameters.

Parameters	Values	Units	Description
Recirculated flow rate (*Rfr*)	300	m^3/h	
Volume of room (*V*)	100	m^3	
Filter efficiency (*Feff*)	90	%	MERV 13 – Enter from **Table 5**
Removal in ducts, Air handler (*Rd*)	10	%	Assuming
Other removal	0	%	Germicidal UV
Measures (*Rot*)			
ACH for additional control measures	3	h^{-1}	$ACH = \frac{Rfr}{V} \cdot MIN(F_{eff} + Rd + Rot)$

Table 6.
ACH (Air Changes Per Hour) for a HEPA filter.

suggests that it can remove at least 90% of the aerosol-sized virus-containing particles. High-efficiency particulate air (HEPA) filters [20] are a type of air filter that can remove up to 99.97% of dust, pollen, mildew, and other airborne particles up to 0.3 microns in size, making them effective at reducing the risk of covid. **Table 6** illustrates an example of ACH calculation for a HEPA filter.

2.7 Fraction of immune people

Vaccination and spontaneous infection are the two main ways to gain immunity [21]. This is most commonly acquired through vaccination, which delivers antigens or attenuated forms of diseases into the body and stimulates an immune response. However, acquiring natural immunity through getting sick and recovering from disease is also possible. Multiple online resources offer accurate information that can be used to calculate the vaccination rate in a given community. Information on vaccination rates can be gathered from various places, including the New York Times' vaccine tracker. Additional country-specific resources, such as the https://lab24.ilsole24ore.com/coronavirus/website, are available. The efficiency of COVID-19 immunizations depends on the vaccine, the period between doses, and the prevalent variants in a given population. The Ref. [22] provides comprehensive information on the percentages of the efficiency of the most effective vaccines for COVID-19, including Pfizer, Moderna, AstraZeneca, and Johnson & Johnson, with an efficiency value of 95.0%, 94.1%, 80.7%, and 66.9% respectively.

As shown in Eq. (6), the number of immune individuals is equal to the product of the number of vaccinated individuals and the efficiency of the vaccine.

$$N_{in} = N_{vac} * V_{eff} \tag{6}$$

where N_{in} is the number of immune people, N_{vac} is the number of people vaccinated and V_{eff} is the percentage of vaccine efficiency.

2.8 CO_2 emission rate

Metabolic production of carbon dioxide (CO_2) by humans is adopted as an indicator to measure ventilation rates in occupied rooms [23]. In this context, individual CO_2 emission rates are determined by analyzing human metabolism, energy-

Activity	M (met)	Range
Dancing aerobic, general	7.3	
Health club exercise classes general	5.0	
Kitchen activity moderate effort	3.3	
Lying or sitting quietly		1.0 to 1.3
Sitting reading, writing, typing	1.3	
Sitting tasks, light effort (e.g., office work)	1.5	
Sitting quietly in religious service	1.3	
Sleeping	0.95	
Standing quietly	1.3	
Standing tasks, light effort (e.g., store clerk, filing)	3.0	
Walking, less than 2 mph, level surface, very slow	2.0	
Walking, 2.8 mph to 3.2 mph, level surface, moderate pace	3.5	

Table 7.
Values of physical activity levels (met).

consuming physical activity, and factors such as ventilation and IAQ. Furthermore, the anthropometric variables of body mass, gender, and age are considered when analyzing the individuals in question.

- Initially, determine the metabolic rate (met) for the activity of interest presented in **Table** 7.
- Subsequently, determine the rate of CO_2 production by taking into account the age, sex, and metabolic activity of the individual.
- If the value of met exceeds 4, as indicated in **Table** 7, it is admitted to use the maximum value in **Table 8** to satisfy the requirement of 4.

Suppose an individual exceeds the threshold of 4. In this case, it is essential to recognize that maintaining such levels of physical activity for extended periods is typically unfeasible for the general population, except for professional athletes.

Based on the outlined rules and tabulated data, the CO_2 emission rate can be computed as follows:

$$CO_2 = (met * N) * \frac{1}{Pr} * \frac{273.15 + T}{273.15} \quad (7)$$

where CO_2 is the emission rate (for all people) and is expressed in L/s; Pr is the Pressure and is expressed in atm; met is the generation rate; N is the number of people present; T is the Temperature, measured in °C.

In addition, CO_2 mixing ratio is given by the following formula:

$$CO_{2AVG_m} = \frac{(CO2 * 3.6)}{\lambda v \ V} * \left(1 - \left(\frac{1}{\lambda v \ D}\right) * \left(1 - e^{-\lambda v \ D}\right)\right) * 1000000 + B_{CO_2} \quad (8)$$

			CO_2 generation rate (L/s)						
	Mean	**Basal**	**Level of physical activity (met)**						
Age	**Body**	**Metabolic**							
	Mass	**Rate (BMR)**							
[years]	**[Kg]**	**[MJ/day]**	**1.0**	**1.2**	**1.4**	**1.6**	**2.0**	**3.0**	**4.0**
Males									
to < 21	77.3	7.77	0.0037	0.0045	0.0053	0.0060	0.0059	0.0113	0.0150
to < 30	84.9	8.24	0.0039	0.0048	0.0056	0.0064	0.0063	0.0120	0.0160
to < 40	87.0	7.83	0.0037	0.0046	0.0053	0,0061	0.0059	0.0114	0.0152
to < 50	90.5	8.00	0.0038	0.0046	0.0054	0.0062	0.0060	0.0116	0.0155
Females									
to < 21	65.9	6.12	0.0029	0.0036	0.0042	0.0047	0.0059	0.0089	0.0119
to < 30	71.9	6.49	0.0031	0.0038	0.0044	0.0050	0.0063	0.0094	0.0126
to < 40	74.8	6.08	0.0029	0.0035	0.0041	0,0047	0.0059	0.0088	0.0118
to < 50	77.1	6.16	0.0029	0.0036	0.0042	0.0048	0.0060	0.0090	0.0119

Table 8.
CO_2 generation rates for ranges of ages and physical activity, at 273°K and 101 KPa.

where CO_{2AVG_m} is the avg. mixing ratio (ppm); CO_2 is the emission rate (for all people) (L/s); λv is the added effects of ventilation $(1/hours)$; D is the duration of exposure $(hours)$; B_{CO_2} is the Background CO_2 Outdoors (ppm) measured in the range between 350 and 450 ppm.

3. Modeling infection risk

3.1 Infection probability

The quantum emission rates exhibit a wide range of variation, depending mainly on the nature of the activities to be performed. Activities with low intensity correspond to lower emission rates. In contrast, loud talking, shouting, and singing correspond to higher emission rates, as highlighted by data reported in **Table 1**. The volumetric breathing rates vary depending on the specific activity being performed, as highlighted by data presented in **Table 2**. Despite the presence of uncertainties in the emission values of SARS-CoV-2 quanta/h, it is feasible to compute estimates of infection risk and draw comparisons regarding the impact of factors such as ventilation, room characteristics, duration of exposure, and the adoption or non-adoption of masks. The estimation of the probability of infection as a function of the duration of exposure time for a set of commonly used environments is depicted in **Figure 1**. People between the ages of 21 and 31 are considered. It is also assumed that at least one person is infected. The study evaluates different quantum emission rates for various settings. For instance, a quantum emission rate of 11.4 quanta/h is assumed to be appropriate for school and office environments, while a rate of 26.3 quanta/h is considered to be suitable for restaurants. Similarly, a quantum emission rate equal to 5.6 quanta/h is considered to be appropriate for libraries, and a rate of 63.1 quanta/h is

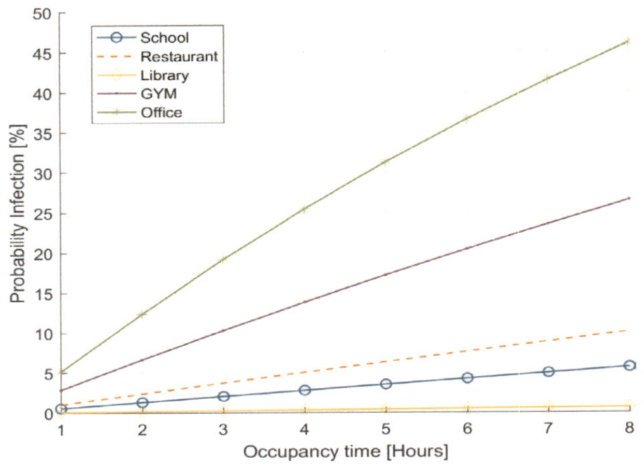

Figure 1.
Relationship between occupancy time and probability of infection in various common environments.

suitable for gyms. The results depicted in **Figure 1** indicate a positive correlation between the duration of exposure and the probability of contracting an infection. Based on a two-hour exposure time, the probability of infection is relatively low, specifically below 5%, for a library accommodating 20 individuals, a school classroom with 21 students, and a restaurant with 78 people. In contrast, in high-intensity activity settings, such as a GYM, the probability of contagion increases more rapidly, due to the increase in the emission of quanta by individuals. There is also a higher probability of infection in office rooms due to their relatively small dimensions and typical occupancy ranging from one to two people.

3.2 The impact of mask-wearing

Figure 2(a) displays the projected correlation between the probability of contracting an infection, the duration of residency, and the number of ACH for

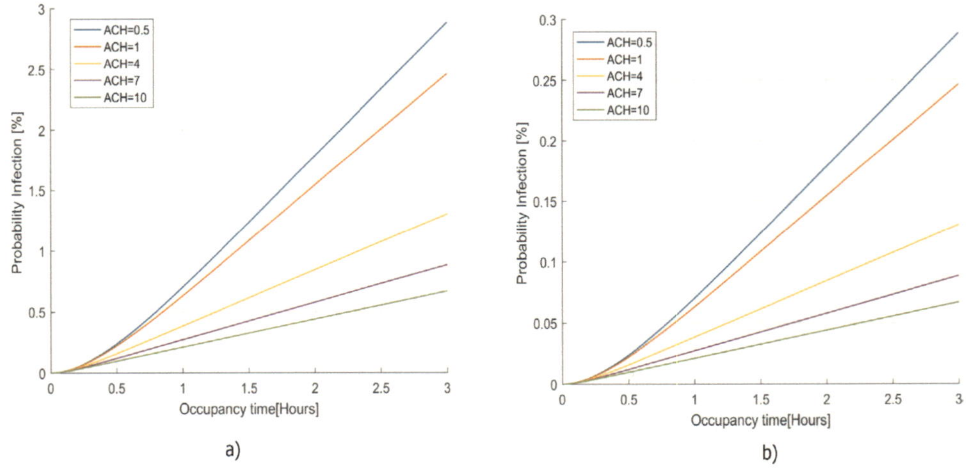

Figure 2.
Probability of infection vs. permanence time, for different ACH values: (a) without a mask (b) with a mask.

individuals wearing a mask. Conversely, **Figure 2(b)** illustrates the same association for individuals not wearing masks. The results suggest that using masks is a significant factor in mitigating the probability of contracting an infection. In the case of an infected individual being present in a given space, the adoption of masks by both the infected and susceptible individuals may reduce the probability of infection to less than 1%, thereby requiring a specific ACH. The results depicted in **Figure 2** indicate that a higher ACH value corresponds to a relatively low probability of infection, with a rate below 5%. Conversely, a decrease in the ACH value is associated with an increase in the probability of infection.

3.3 CO_2 generation rate

The rate of CO_2 generation is evaluated for different environments featuring activities of varying intensities, and the results are illustrated in **Figure 3**. The correlation between the CO_2 concentration and the number of quanta required by an individual in an enclosed space is observed. Similarly, CO_2 concentrations tend to considerably increase over time, depending on the nature of the task and the number of individuals present.

The reference values for the concentration of CO_2, as defined in the literature [24], are considered as threshold values. A CO_2 concentration of less than 450 ppm is typically recommended for outdoor air, while indoor environments during pandemic situations are advised to maintain a reference value of 800 ppm for CO_2. However, it is also acceptable for indoor environments that CO_2 levels reach up to 1000 ppm. For levels greater than 1000 ppm, it is recommended that the duration of exposure should not surpass 8 hours. The sustained elevation of CO_2 levels over time can indicate insufficient ventilation relative to the number of occupants and the activities to be conducted.

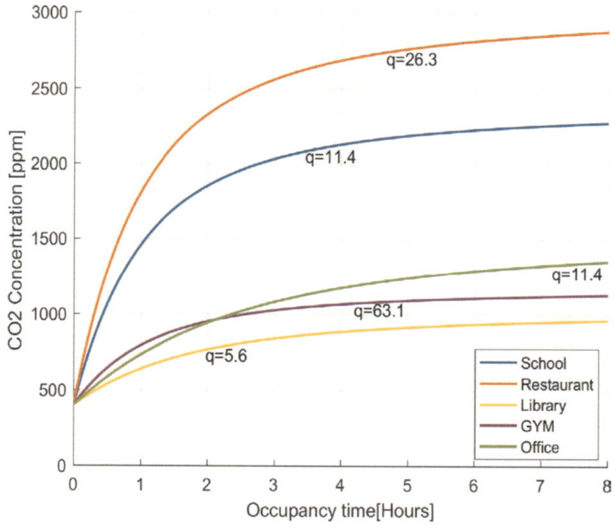

Figure 3.
Relationship between carbon dioxide (CO_2) concentration and permanence time across various indoor environments.

3.4 Social distance to prevent COVID-19

Implementing social distancing measures is a crucial consideration to mitigate the transmission of viral infections within enclosed environments. The distance index can be derived through theoretical analysis of droplet dispersion and transmission resulting from human respiratory activities. According to the research conducted in [25], it is suggested that a distance of 1.6–3.0 m (5.2–9.8 ft) may be assumed as a safe criterion in terms of aerosol transmission from large exhaled droplets during speech. However, when accounting for all droplets in still air, the safe social distance can be extended up to 8.2 m (26 ft). The adoption of Bluetooth technology is suggested to measure social distancing using the received signal intensity indicator (RSSI). The current approach exploits the functionalities of Bluetooth-enabled devices to determine proximity among individuals and mitigate the transmission of infectious diseases. Observing RSSI values of packets received from proximate devices makes it feasible to approximate the distance between individuals. As the distance between devices increases, RSSI values tend to decrease, thereby establishing a negative correlation between signal strength and proximity, as seen in **Figure 4(a)**. In **Figure 4(b)**, the application of curve fitting to experimental data is considered, wherein distances ranging from 0 to 7 meters between devices are assumed. This approach is applied to accurately correlate the distance and the RSSI-measured values. Eq. (9) represents the approximation model that depicts the correlation between the distance y measured in meters and the RSSI (dBm) value, namely:

$$y = a1 * sin\,(b1 * x + c1) \;+\; a2 * sin\,(b2 * x + c2) \;+\; a3 * sin\,(b3 * x + c3) \qquad (9)$$

where: a1 = 23.57; b1 = 0.005066; c1 = −2.92; a2 = 1.168; b2 = 0.09591; c2 = −1.643; a3 = 0.42; b3 = 0.2547; c3 = −3.135.

4. Experimental setup

The previously outlined model applied in this section to assess the probability of COVID-19 transmission is depicted in **Figure 5**, which incorporates all relevant parameters and equations. An application with specific functionalities is developed

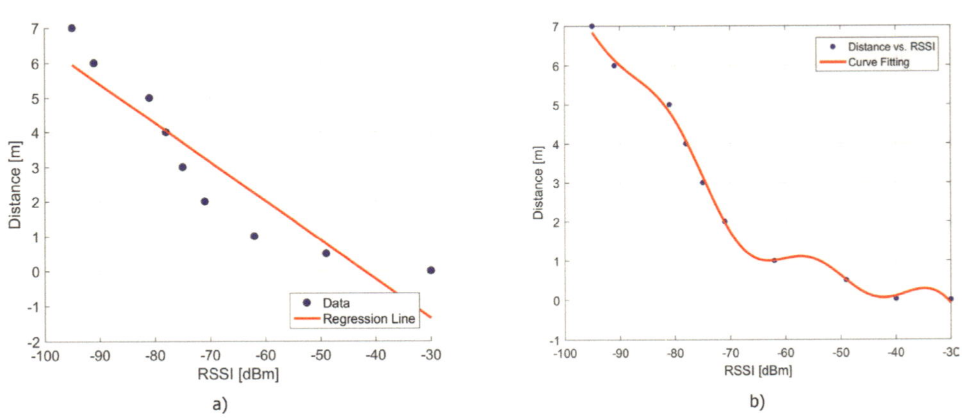

Figure 4.
Distance measurements: (a) Relationship between RSSI and Distance; (b) Curve fitting to experimental data.

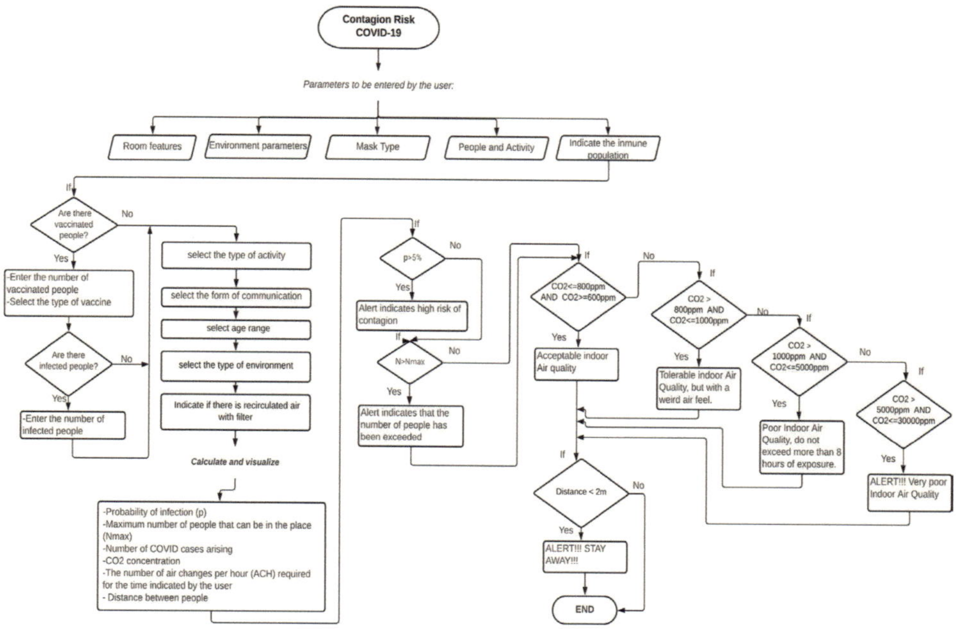

Figure 5.
Flowchart for estimating the COVID-19 contagion risk.

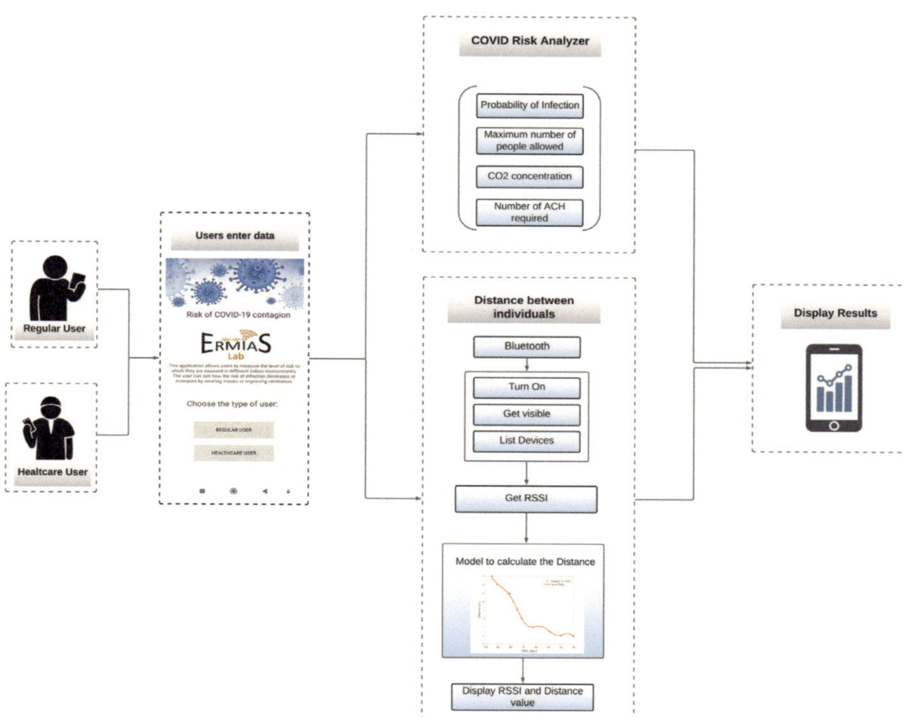

Figure 6.
Application architecture.

using Android Studio. An extensive representation of the architecture for the proposed system is illustrated in **Figure 6**. In particular, the system allows to choose between two distinct user categories: Regular User and Healthcare User. Three distinct blocks exist, namely the user input parameters, the risk analyzer, and the measurement of interpersonal distance. They are described in detail as follows.

1. Users enter data: in this block, the application requests the user to submit specific parameters related to the prediction of COVID-19, on the basis of the specific indoor environment. In the case of a regular user, an interface is presented that allows him to input known parameters and easily comprehend and navigate the application. In this instance, specific parameters are omitted, including the dimensions of the location (standard dimensions are considered for each environment), the number of infected individuals, the number of vaccinated individuals, the number of individuals who are already infected, and the type of vaccine. In contrast, healthcare users must input these more specific parameters to obtain more precise results. The information gathered by this block is sent to the Covid Risk Analyzer block.

2. Covid Risk Analyzer block: this block receives input from the first block; it processes data and computes the probability of contracting COVID-19, the maximum occupancy of a given space, the concentration of CO_2, and the necessary ACH required for the environment. Furthermore, this block enables the application to display alert notifications to the user under certain circumstances, namely:

    - when the probability of contagion is high;

    - once the maximum admitted capacity of individuals is overcome;

    - when the air quality is suboptimal, specifically when the concentration of CO_2 exceeds 1000 ppm.

3. Distance between individuals: the distance monitoring functionality is devised to collect Bluetooth data, including the approximated RSSI metric, and assess the proximity between close devices. In order to make use of this functionality, users must enable their Bluetooth connectivity. Upon activation, the system consistently monitors the distances among its users. If the proximity between individuals is measured under two meters, a notification alert is promptly generated to inform the user. **Figure 7** depicts the visual representation of the block and its functionality. The above feature is essential to promote social distancing and ensure user's safety.

5. Results and discussion

This section illustrates some relevant assessment cases of the application outlined in Ref. [26], by emphasizing critical scenarios essential for evaluating the probability of infection and quantifying distance between individuals. The study is limited to participants aged from 21 to 31 years, as this age group is the most prominent

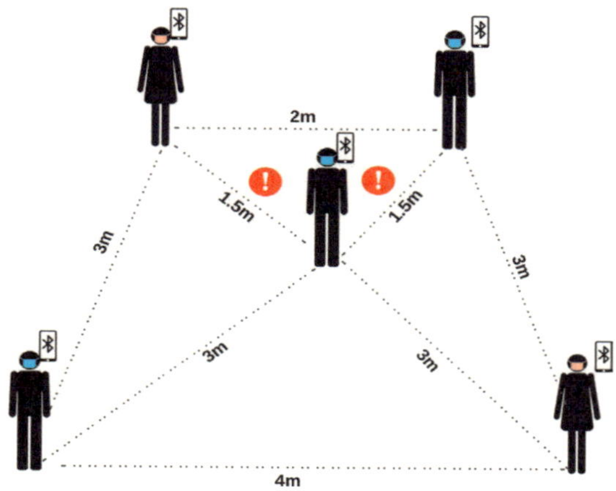

Figure 7.
Schematic representation of social distancing.

demographic for the selected scenarios in our analysis. However, users are not limited to this age range, and they can confidently choose any other age group.

5.1 Scenario 1-classroom

Scenario 1 involves a parameter analysis in a classroom with 61.2 m^2 in the area and 4.9 meters in height. A total number of 22 individuals are considered, with ages ranging from 21 to 31 years. It is important to note that in this scenario all individuals present wear facial masks, and the lesson time is scheduled to last two hours. The involved activity is characterized by light intensity and speaking. Significantly, this particular scenario considers the existence of ventilation with filters. Under these conditions, it is observed from **Figure 8** that the probability of contracting an infection stands at 1.19%. Additionally, the concentration of CO_2 is measured at 927, indicating a tolerable level of indoor air quality.

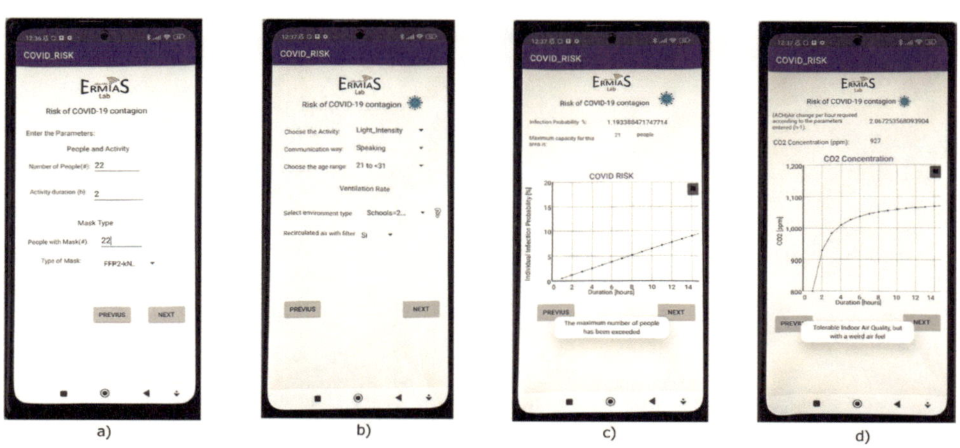

Figure 8.
Evaluation of Scenario 2-Classroom.

5.2 Scenario 2-restaurant

In scenario 2, the case of a restaurant with an area of 111.48 m^2 and a height of 7.31 m is considered. The number of present people is equal to 78, with ages ranging from 21 to 31 years. They spent a total of 2 hours in the designated location. In this scenario, individuals do not wear masks, while filters facilitate ventilation. Moderate intensity and speaking are assumed for this activity. **Figure 9** illustrates that the probability of infection under these conditions equals 2.08%, and the maximum allowed occupancy limit is not exceeded. However, by examining a significant population, the CO_2 concentration during this period assumes the high value of 1997, indicating poor indoor air quality. So, it is advisable to reduce the permanence duration.

5.3 Scenario 3-sports and entertainment

The present scenario involves the analysis of a GYM, which encompasses an area of 371.61 m^2 and a height of 3.66 meters. There are 20 individuals in attendance, ranging in age from 21 to 31 years, and anyone wearing masks. The activity lasts two hours, characterized by high-intensity and loud speaking activity. Furthermore, the consideration of filter ventilation is included in this particular scenario. The analysis depicted in **Figure 10** indicates a significant infection probability of 13.16%. Although the maximum occupancy limit has not been surpassed and the concentration of CO_2 is within acceptable levels, limiting the duration of stay is recommended due to the high intensity of the activity being conducted in this environment.

5.4 Distance between individuals

Figure 11 illustrates the operational capabilities of the application, demonstrating its capacity to exhibit a complete list of nearby devices, accompanied by their respective IP addresses, Indicator RSSI measurements, and approximated distances. If the distance of two meters between individuals is exceeded, a notification is sent to the user's smartphone. This alert reminds individuals to modify their physical proximity and comply with the recommended social distancing protocols.

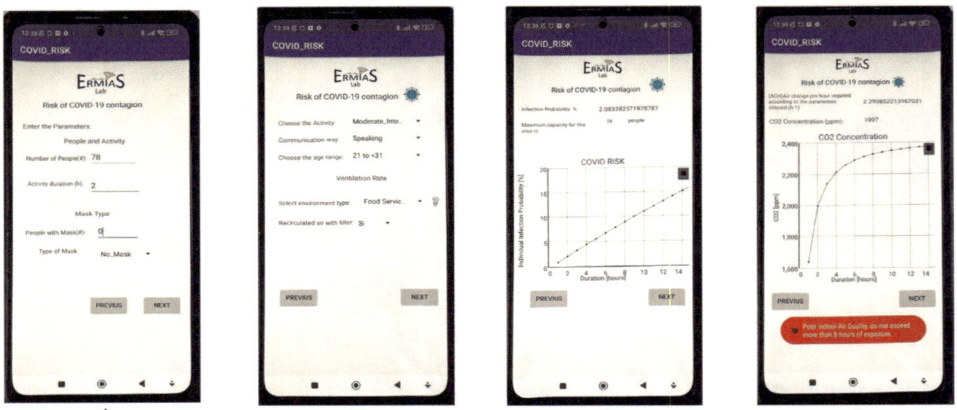

Figure 9.
Evaluation of scenario 3-restaurant.

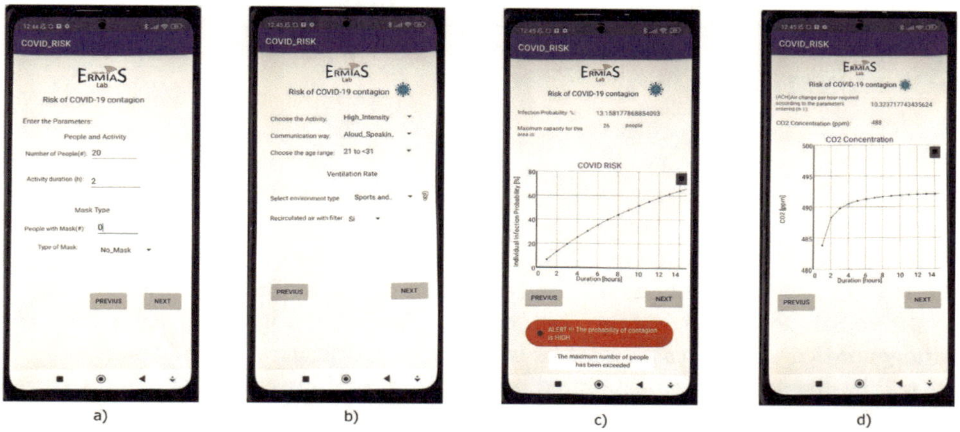

Figure 10.
Evaluation of scenario 6-sports and entertainment.

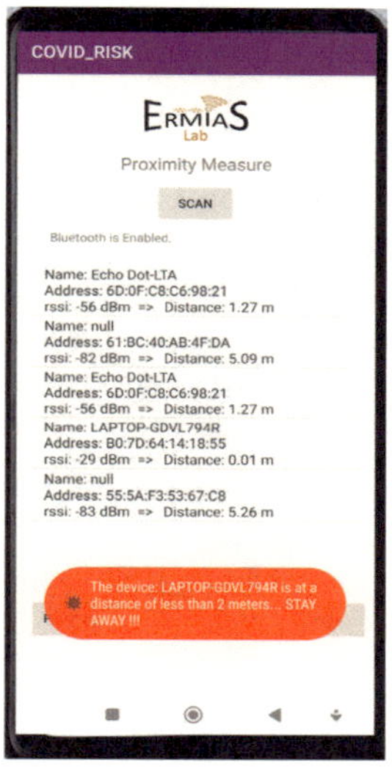

Figure 11.
Evaluation of distance between individuals.

6. Conclusions

An accurate model based on the Wells-Riley probabilistic methodology has been presented and discussed in this chapter to predict COVID-19 transmission in indoor environments. The basic theory of the proposed model has been deeply illustrated, by emphasizing its accuracy in assessing the transmission probability. In order to facilitate continuous monitoring of COVID-19 infection risk in indoor environments, the model has been incorporated into an innovative Android application that also computes the distance between individuals. The results obtained during the assessment stage of the model indicate that a combination of factors should be adopted accurately to manage the transmission of COVID-19 in indoor environments effectively. A proper approach should in fact include factors such as optimal ventilation, adherence to mask-wearing protocols, respect for occupancy limits, the minimization of time spent in closed spaces, and the maintenance of a minimum physical distance of two meters between individuals.

Acknowledgements

This research was funded by PNRR project Age-IT (Ageing well in an ageing society) - Conseguenze e sfide dell'invecchiamento, and by Progetto POS (Piano Operativo Salute) 2021 dal titolo "Radioamica – Open Network per la RADIOmica/ rAdiogenoMIca Cooperativa basata su intelligenza artificiale" - Traiettoria 2 "E-Health, diagnostica avanzata, medical devices e mini invasività", Azione 2.1 "Creazione di una rete nazionale per le malattie ad alto impatto" del Piano Sviluppo e Coesione Salute - FSC 2014-2020.

Abbreviations

IAQ	Indoor Air Quality
ACH	Air Changes per Hour
MERV	Minimum Efficiency Reporting Values
PSE	Particle Size Efficiency
HEPA	High-Efficiency Particulate Air
atm	Standard Atmosphere
RSSI	Received Signal Intensity Indicator
ppm	Parts per Million
IP	Internet Protocol

Author details

Sandra Costanzo[1,2,3,4]*† and Alexandra Flores[1]†

1 Department of Computer Engineering, Modeling, Electronics and Systems, University of Calabria, Rende, Italy

2 ICEmB (Inter-University National Research Center on Interactions between Electromagnetic Fields and Biosystems), Genoa, Italy

3 National Research Council of Italy (CNR), Institute for Electromagnetic Sensing of the Environment (IREA), Naples, Italy

4 CNIT (National Interuniversity Consortium for Telecommunications), Parma, Italy

*Address all correspondence to: costanzo@dimes.unical.it

† These authors contributed equally.

References

[1] World Health Organization. Coronavirus Disease (COVID-19). 2023. Available from: https://www.who.int/health-topics/coronavirus#tab=tab_1 [Accessed: June 18, 2023]

[2] World Health Organization. Q&A: How is COVID-19 Transmitted?. 2020. Available from: https://www.who.int/vietnam/news/detail/14-07-2020-q-a-how-is-covid-19-transmitted [Accessed: June 18, 2023]

[3] Pan Y, Zhang D, Yang P, Poon LLM, Wang Q. Viral load of SARSCoV-2 in clinical samples. The Lancet Infectious Diseases. 2020;**20**:411-412. DOI: 10.1016/S1473-3099(20)30113-4

[4] Wölfel R, Corman VM, Guggemos W, et al. Virological assessment of hospitalized patients with COVID-2019. Nature. 2020;**581**:465-469. DOI: 10.1101/2020.03.05.20030502

[5] Miller SL, Nazaroff WW, Jimenez JL, Boerstra A, Buonanno G, Dancer SJ, et al. Transmission of SARS-CoV-2 by inhalation of respiratory aerosol in the Skagit Valley Chorale superspreading event. Indoor Air. 2021;**31**(2):314-323

[6] Qian H, Miao T, Liu L, Zheng X, Luo D, Li Y. Indoor transmission of SARS-CoV-2. medRxiv. 2020;**31**(3):639-645. DOI: 10.1101/2020.04.04.20053058

[7] Nishiura H, Oshitani H, Kobayashi T, et al. Closed environments facilitate secondary transmission of coronavirus disease 2019 (COVID-19). medRxiv. 2020;**2020**(2). DOI: 10.1101/2020.02.28.20029272

[8] Li X, Lester D, Rosengarten G, Aboltins C, Patel M, Cole I. A spatiotemporally resolved infection risk model for airborne transmission of COVID-19 variants in indoor spaces. Science of the Total Environment. 2022; **812**:152592

[9] Zhao X, Liu S, Yin Y, Zhang T, Chen Q. Airborne transmission of COVID-19 virus in enclosed spaces: An overview of research methods. Indoor Air. 2022;**32**(6):e13056

[10] Wells WF. Airborne Contagion and Air Hygiene: An Ecological Study of Droplet Infections. New York: Commonwealth Fund; 1955

[11] Riley RL, Nardell EA. Clearing the air: The theory and application of ultraviolet air disinfection: Reply. American Review of Respiratory Disease. 1990;**142**(5):1233-1234

[12] Beca-Martínez MT, Romay-Barja M, Falcón-Romero M, Rodríguez Blázquez C, Benito-Llanes A, Forjaz MJ. Compliance with the main preventive measures of COVID-19 in Spain: The role of knowledge, attitudes, practices, and risk perception. Transboundary and Emerging Diseases. 2022;**69**(4):e871-e882

[13] Kurnitski J. Ventilation rate and room size effects on infection risk of COVID-19. The REHVA European HVAC Journal. 2020;**57**(5):26-31

[14] Jiménez J. COVID-19 Airborne Transmission Tool Available. Cooperative Institute for Research in Environmental Sciences (CIRES) 2020. Available from: https://cires.colorado.edu/news/covid-19-airborne-transmission-tool-available [Accessed: June 18, 2023]

[15] Buonanno G, Stabile L, Morawska L. Estimation of airborne viral emission:

Quanta emission rate of SARS-CoV-2 for infection risk assessment. Environment International. 2020;**141**:105794

[16] Morawska L, Johnson GR, Ristovski ZD, Hargreaves M, Mengersen K, Corbett S, et al. Size distribution and sites of origin of droplets expelled from the human respiratory tract during expiratory activities. Journal of Aerosol Science. 2009;**40**(3):256-269

[17] ANSI/ASHRAE Standard 62.1-2013. Ventilation for Acceptable Indoor Air Quality. 2013. Available from: https://www.ashrae.org/permissions [Accessed: June 18, 2023]

[18] Dabisch P, Schuit M, Herzog A, Beck K, Wood S, Krause M, et al. The influence of temperature, humidity, and simulated sunlight on the infectivity of SARS-CoV-2 in aerosols. Aerosol Science and Technology. 2021;**55**(2): 142-153

[19] NAFA. Understanding MERV — NAFA User's Guide to ANSI/ASHRAE 52.2. 2018. Available from: https://www.nafahq.org/understanding-merv-nafa-users-guide-to-ansi-ashrae-52-2/ [Accessed: June 18, 2023]

[20] Zhai Z, Li H, Bahl R, Trace K. Application of portable air purifiers for mitigating COVID-19 in large public spaces. Buildings. 2021;**11**(8):329

[21] Speiser DE, Bachmann MF. COVID-19: Mechanisms of vaccination and immunity. Vaccine. 2020;**8**(3):404

[22] Mohammed I, Nauman A, Paul P, Ganesan S, Chen KH, Jalil SM, et al. The efficacy and effectiveness of the COVID-19 vaccines in reducing infection, severity, hospitalization, and mortality: A systematic review. Human Vaccines & Immunotherapeutics. 2022;**18**(1):2027160

[23] Qi MW, Li XF, Weschler LB, Sundell J. CO2 generation rate in Chinese people. Indoor Air. 2014;**24**(6):559-566

[24] Peng Z, Jimenez JL. Exhaled CO2 as a COVID-19 infection risk proxy for different indoor environments and activities. Environmental Science & Technology Letters. 2021;**8**(5):392-397. DOI: 10.1021/acs.estlett.1c00183

[25] Sun C, Zhai Z. The efficacy of social distance and ventilation effectiveness in preventing COVID-19 transmission. Sustainable Cities and Society. 2020;**62**: 102390

[26] Costanzo S, Flores A. COVID-19 contagion risk estimation model for indoor environments. Sensors. 2022; **22**(19):7668

Chapter 3

Indoor Environmental Quality Study for Higher Education Buildings

Mukhtar Maigari, Changfeng Fu, Jie Deng and Ali Bahadori-Jahromi

Abstract

Indoor environmental quality (IEQ) in school buildings has been concerned widely for many years, whilst research into the IEQ issues in higher education (HE) buildings has been overlooked to some extent. This chapter presents an experimental study of the IEQ issues in two typical HE buildings in London using the post-occupancy evaluation (POE) methods. Various aspects of the IEQ have been considered in terms of human comfort in buildings, including indoor air quality, noise level, lighting, occupants' perception, and so on. IEQ data have been collected using various IEQ meters and data loggers, as well as questionnaire surveys taken by the respondents. The results of the study reveal important findings. In terms of thermal comfort, several spaces were found to exceed the recommended temperature limit of 25°C. The data on indoor air quality indicated that rooms, particularly those with natural ventilation, such as the architectural studio, significantly exceeded the recommended CO_2 limit of 1500 ppm. Moreover, the survey feedback collected from the building occupants aligned with the IEQ data, particularly in the area of thermal comfort. The respondents' feedback provided valuable insights into their experiences and perceptions of the indoor environment, further reinforcing the findings obtained from the objective IEQ measurements. The work also discusses recommendations and possible actions to improve the IEQ in HE buildings.

Keywords: indoor environmental quality (IEQ), higher education (HE) buildings, post occupancy evaluation (POE), indoor air quality (IAQ), IEQ standards and guidelines

1. Introduction

Indoor environmental quality (IEQ) is a critical aspect of the built environment that has a significant impact on the health, well-being, and productivity of building occupants. The quality of indoor air, thermal comfort, lighting, acoustics, and ergonomics are some of the factors that determine the IEQ of an infrastructure in the built environment [1]. In higher education (HE) buildings, such as classrooms, libraries, laboratories, and lecture halls, etc., the importance of IEQ cannot be overemphasized. Previous studies and research on the state of IEQ in educational institutions have predominantly centred around primary and secondary schools [2, 3]. The nature of higher educational buildings with complex spaces and structures consisting of various rooms such as laboratories, lecture theaters, PC rooms etc., has possibly added to the limited studies and research on the state of IEQ in HE buildings. As such, there has been heightened interest in understanding the IEQ state of HE buildings especially following the COVID-19 pandemic which has greatly increased awareness towards IEQ especially in public spaces such as educational institutions. This has led to the development of various strategies and tools for assessing and improving IEQ.

One of the most effective strategies for assessing IEQ in higher education buildings is the use of post-occupancy evaluation (POE) methods. POE involves the systematic evaluation of a building's performance after it has been occupied by its intended user [4]. POE methods provide a comprehensive assessment of the IEQ of a building by combining data from various sources, such as physical measurements, occupant feedback, and building performance data. However, despite the potential benefits of using POE methods to assess IEQ in higher education buildings, there is still a lack of awareness and understanding of these methods amongst building managers and stakeholders. As such, this article aims to address this apparent gap, coupled with the lack of IEQ awareness and understanding in higher education buildings through the following objectives:

- To gain comprehensive understanding on the current state of knowledge on IEQ and POE by reviewing relevant literature including peer reviewed journal articles, books, and reports.

- To identify the existing IEQ situations in HE buildings and the current means for IEQ monitoring and control through case studies.

- To gain invaluable insights from students regarding their perception of IEQ with a questionnaire survey.

- To highlight issues with IEQ in HE buildings based on the findings in the literature review, case study, and student feedback as well as provide recommendations for how to mitigate such issues.

2. Background

2.1 Indoor environmental quality (IEQ)

The indoor environment constitutes the different kinds of indoor spaces available within built assets, such as residential buildings, offices, schools, and hospitals. The state of the indoor environment has been a prominent research area and industry interest even before the COVID-19 pandemic. Improving the quality of life of building occupants, increasing work performance or simply in a bid to make a building more sustainable are few reasons why research into IEQ has gained lots of traction in the AEC industry [5]. IEQ also refers to "the quality of a building's environment in relation to the health and wellbeing of those who occupy space within it" [6]. The IEQ is linked to indoor human comfort which is usually assessed from four aspects: thermal, respiratory, visual, and acoustic comfort. Respiratory comfort is generally

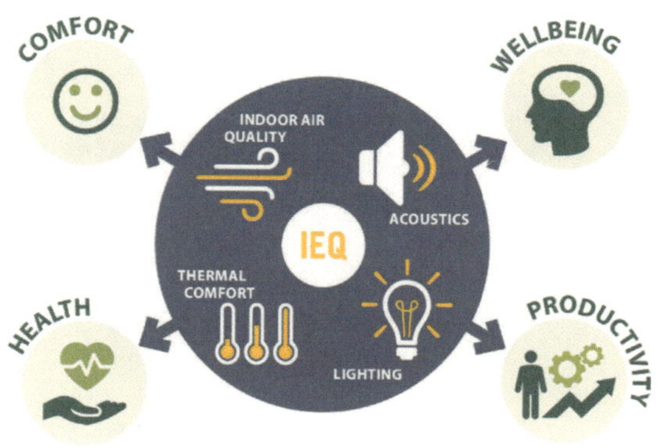

Figure 1.
IEQ embodiment [7].

expressed as indoor air quality (IAQ). These four factors all combine to affect the comfort, health, well-being, and productivity of building occupants as depicted in **Figure 1**.

Indoor environmental quality (IEQ) has a significant impact on the health, comfort, and productivity of building occupants, especially in educational buildings where students and staff spend a considerable amount of time. According to studies, poor IEQ can lead to discomfort, respiratory problems, allergies, and other health issues [8]. In higher education buildings, poor IEQ can also affect students' academic performance, attendance, and retention [9].

Assessing IEQ in higher education buildings is essential to ensure that the indoor environment is healthy, comfortable, and conducive to learning. The conventional method of evaluating IEQ in buildings involves using post-occupancy evaluation (POE) methods to collect data from building occupants after they have used the space for a while. POE involves gathering feedback from occupants through surveys, interviews, and other data collection methods to evaluate their experience of the indoor environment.

2.2 IEQ factors

2.2.1 Thermal comfort (TC)

The American Society of Heating, Refrigerating, and Air-Conditioning Engineers (ASHRAE) and International Organization for Standardization (ISO) defined thermal comfort as "that condition of mind that expresses satisfaction with thermal environment" for thermal comfort [10, 11]. This definition is influenced by the significant contributions of Professor Povl Ole Fanger in the field of thermal comfort. His work, including his dissertation and book titled "Thermal Comfort," introduced a novel relationship between environmental physical parameters, human physiological parameters, and comfort perception [12]. This led to the development of the prediction mean vote (PMV) and prediction percentage dissatisfied (PPD), often known as the chamber model [13]. According to the Health and Safety Executive body (HSE), thermal comfort is affected by a combination of environmental and personal factors which are represented in **Figure 2**.

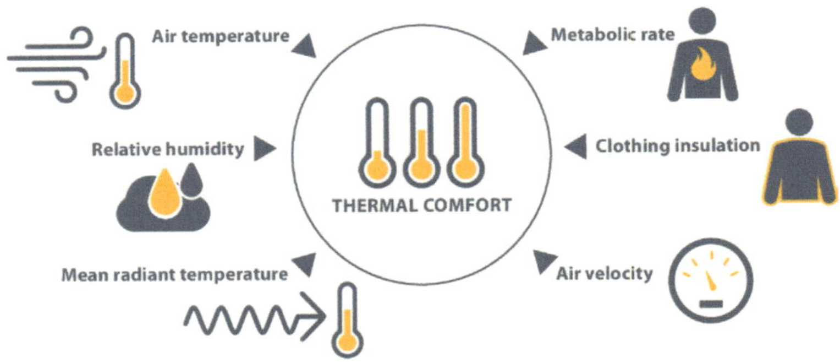

Figure 2.
Factors that affect thermal comfort [14].

2.2.2 Indoor air quality (IAQ)

The rapid progress in technology in recent times has ushered in the digital era, causing a significant shift in human behavior towards spending more time indoors. Whether it is office workers, students in educational institutions, patients in hospitals, or individuals in their homes, studies have shown that approximately 90% of people's time is spent in indoor environments [15]. This reality underscores the importance of maintaining the quality of the indoor environment, particularly the indoor air quality (IAQ).

ASHRAE, in their indoor air quality guide [16], defined IAQ as the "air in which there are no known contaminants at harmful concentrations as determined by cognizant authorities and with which a substantial majority (80% or more) of the people exposed do not express dissatisfaction" [16]. Ensuring good IAQ is critical, and [17] outlined three key reasons for this:

- Indoor air serves as the interaction medium amongst weather conditions, people, and buildings.
- The physical, biological, and chemical characteristics of indoor air directly influence the health and well-being of building occupants.
- Given the straightforward nature of indoor air, IAQ can be clearly defined and managed to meet required standards.

Various studies have identified factors that influence IAQ, including temperature (°C), relative humidity (RH %), and pollutants (chemical, biological, and physical) [18]. Amongst these factors, pollutants play a significant role, with elements such as carbon dioxide (CO_2), nitrogen dioxide (NO_2), volatile organic compounds (VOCs), and viruses being of particular concern.

2.2.3 Acoustic comfort (AC)

Another important IEQ parameter is acoustic comfort (AC) which centres around "noise": an "unwanted sound" which when present in the built environment, affects concentrations, interferes with activities, prevents speech communication, and if in high levels can impair hearing significantly. Thus, the capacity of a building to provide a suitable acoustic environment and protection against noise in line with the necessary acoustic requirements is the acoustic comfort of a building [14]. Noise, being a form of pollution, evidently translates to the crucial impact acoustic comfort has in the health & well-being, productivity as well as communication of a building's occupants [19]. The acoustic environment's comfort is typically influenced by various factors, including the acoustic properties (sound absorption, transmission, and reflection) of the indoor space, the geometry and volume of the indoor area, the transmission of airborne noise, impact noise as well as noise from internal and external sources, such as background noise [20].

2.2.4 Visual comfort

Visual comfort is an IEQ factor that involves lighting which could either daylighting or artificial lighting [21]. It is an important element of the overall IEQ in education buildings with studies ascertaining it to be a major contributor in the creation of an optimum learning environment [22]. The European Standard (EN 12665, [23]) defined visual comfort as "the subjective visual wellbeing condition induced by the uminous environment" and this definition clearly indicates a psychological element to the overall perception of visual comfort by individuals [24]. It is achieved when the lighting quality and quantity, occupant perception, and environmental quality of view are in a good balance [25]. Lighting quality is a measure of the light's brightness and color, whereas lighting quantity involves the illumination levels and output [26].

2.3 IEQ standards and guidelines for HE buildings

Indoor environmental qualities, like all aspects of sustainability, are mostly governed by certain standards. As established, IEQ is characterized by four environmental factors which are thermal comfort, indoor air quality, acoustic comfort, and visual comfort. These four factors are different with different means of measurements and recording and thus, requirements for each IEQ factor are expected to be different leading to the emergence of different standards for most IEQ factors. In buildings generally, the evaluation and design of the indoor environment are governed by national and international standards. These standards provide guidelines in the specification of acceptable indoor environmental conditions for occupants [27]. They are highlighted as follows.

2.3.1 Baseline designs for schools

These were developed by the education funding agency (EFA) in response to a recommendation in the Review of Education Capital in April 2011. The review called for a suite of standardized drawings and specifications that could be applied across a wide range of educational facilities [28]. The designs provide a light, bright, and airy learning environment for students and teachers. They were drawn up with the advice of environmental, architectural, and teaching experts to address problems such as dark corridors, poor ventilation, and inadequate classrooms, and to make the very best use of space [28]. **Table 1** provides an overview of the published guidelines on various aspects of design and construction:

Publication	Overview
Environmental services strategy	This outlines criteria for indoor air quality, lighting, temperature control, and energy consumption. The document also provides guidance on ventilation strategies for both primary & secondary schools and sets operational targets for energy consumption
Structural strategies	This document provides guidance on the structural design of school buildings
Circulation models	This document provides guidance on the design of circulation spaces within school buildings
Access and inclusion	This document provides guidance on ensuring accessibility and inclusivity in the design of school buildings
Daylight strategy	This document provides guidance on ensuring adequate daylight in school buildings
Acoustic performance standards	This document provides guidance on the acoustic design of school buildings

Table 1.
EFA baseline design guidance document.

2.3.2 Chartered Institution of Building Services Engineers (CIBSE)

CIBSE stands for the Chartered Institution of Building Services Engineers. It is a professional body in the UK for building services engineers, encompassing a wide range of disciplines such as heating, ventilation, air conditioning, lighting, and plumbing. CIBSE was founded in 1976, and its main aim is to promote the science, art, and practice of building services engineering, as well as to promote the efficient use of energy in buildings. It achieves this through various means such as the publication of technical guidance and codes of practice, organizing seminars and conferences, and providing education and training for building services engineers **Table 2** presents an overview of the published CIBSE guides:

CIBSE Guide	Title
CIBSE Guide A	Environmental design: provides a comprehensive overview of environmental design in buildings, including principles of thermal comfort, indoor air quality, lighting, and acoustics
CIBSE Guide B	Heating, ventilation, air conditioning, and refrigeration: provides detailed information on HVAC systems, including design principles, load calculation, system selection, and control strategies
CIBSE TM40	Health and wellbeing in building services: provides guidance on how to design, operate, and maintain building services systems to support the health and well-being of occupants and covers indoor air quality, thermal comfort, lighting, acoustics, and other factors that affect occupant health and productivity

Table 2.
CIBSE guidance documents.

2.3.3 International Organization for Standardization (ISO)

The ISO is an organization consisting of 163 national standard bodies headquartered in Geneva (OECD/ISO, 2016—dependent). Established in 1947, the

ISO Standard	Title
ISO 17772-1:2017	Energy performance of buildings: indoor environmental quality
ISO 7730	Ergonomics of the thermal environment: PMV and PPD indices
ISO 16814:2008	Building environment design: indoor air quality
ISO 10551	Ergonomics of thermal environment on subjective assessment methods
ISO 9920	Ergonomics of thermal environment on clothing insulation

Table 3.
ISO relevant IEQ standards.

ISO covers a range of areas in the field of engineering, business, health, technology, computing, and others [13]. Its main goals are to provide global solutions to worldwide challenges and support innovation by developing "voluntary, consensus-based, market relevant international standards" (OECD/ISO, 2016—dependent). **Table 3** presents the relevant IEQ standards published by the ISO.

2.3.4 European Standard (EN)—CEN

The European Standard (EN—European Norms) are sets of standards developed and adopted by the three European Standard Organizations: The European Committee for Standardization (CEN), the European Committee for Electrotechnical Standardization (CENELEC), and the European Telecommunications Standards Institute (ETSI) (European Commission [29]). It comprises more than 800 member organizations worldwide constituting research entities, private companies, academia, and government organizations [30]. **Table 4** presents the relevant IEQ standards published by CEN.

2.3.5 The American Society of Heating, Refrigerating and Air-conditioning Engineers (ASHRAE)

The American Society of Heating, Refrigerating and Air-conditioning Engineers (ASHRAE) is a global society established in 1894 geared towards the advancement of human well-being through sustainable technologies for the built environment [31]. By fostering innovation and disseminating knowledge, it plays a crucial role in shaping the practices and standards related to heating, ventilation, air conditioning, and refrigeration systems, contributing to the improvement of environmental sustainability and human comfort in buildings.

2.4 IEQ indicator setpoints

The IEQ setpoints are a set of standards and guidelines that define the minimum and maximum acceptable levels of environmental parameters for optimal occupant

European Standard	Title
EN 15251:2007	Indoor environmental input parameters for design and assessment
EN 16798:2019	Energy performance of buildings: ventilation for buildings

Table 4.
European Standard (EN) relevant IEQ documents.

Standard	Temperature (°C)	Relative humidity (%RH)	Carbon dioxide (ppm)	PM2.5 (μg/m³)	TVOC (μg/m³)	HCHO (μg/m³)	Illuminance (lux)	Background noise (dB)
CIBSE A	19–21	40–70%	≤1500	•	≤300	•	300; 500	35
BB 101	20–25	•	1500	25 (1 yr)	≤300 (8 hr)	•	•	•
BB 93	•	•	•	•	•	•		40–45
EN 12464-1:2021	•	•	•	•	•	•	500	•
UK Gov. (24-h mean)	•	•	•	10	•	•	•	•
UK Gov. (Annual)	•	•	•	25	•	•	•	•
UK Gov. (30 min)	•	•	•	•	•	≤100	•	•

Table 5.
IEQ setpoints from various published standards and guidelines.

comfort and well-being. They cover several key factors including thermal, visual, acoustic comfort, and indoor air quality (IAQ). For thermal comfort, the setpoints provide guidelines for temperature and humidity levels that ensure occupant comfort and productivity. For visual comfort, the setpoints provide guidelines for lighting levels and glare control to ensure occupant visual well-being and productivity.

Acoustic comfort setpoints provide guidelines for acceptable noise levels, sound transmission, and reverberation times to ensure occupant acoustic well-being and productivity. Finally, IAQ setpoints provide guidelines for acceptable levels of air pollutants, carbon dioxide, and relative humidity to ensure occupant respiratory health and well-being.

Table 5 presents HE buildings relevant IEQ setpoints published by various organizations such as the Building Research Establishment (BRE), CIBSE and the UK government, which are regularly reviewed and updated to reflect new research and standards. The adherence to IEQ setpoints can result in improved occupant comfort, productivity, and well-being, as well as reduced energy costs and environmental impact.

3. Research methods

3.1 Post-occupancy evaluation (POE) methods

Post-occupancy evaluation (POE) is a method for evaluating the performance of a building after occupancy. POE is an essential tool for assessing the indoor environmental quality (IEQ) in higher education buildings. It involves collecting and analyzing data on the performance of a building's systems, such as HVAC, lighting, acoustics, and thermal comfort, to determine their effectiveness and efficiency. The purpose of a POE is to identify the strengths and weaknesses of a building's design and operation, and to identify opportunities for improvement [32].

POE has been used for decades as a way to evaluate the effectiveness of building design and operation in terms of IEQ. According to a study by Fisk [33], POE has been used in a variety of settings, including office buildings, schools, and hospitals. In the context of higher education buildings, POE can be particularly useful in assessing the quality of the learning environment, which is a critical factor in the success of students.

For this study, the primary POE methods used are physical monitoring and questionnaire surveying. Physical monitoring, often being the more objective approach of the two POE methods, involves ascertaining the actual conditions of the building [34]. It may include the use of various measurement instruments and sensors to monitor IEQ indicators such as temperature, humidity, air quality, and lighting levels. Through physical monitoring, researchers and building management teams can gain insights into the actual performance of the building and identify areas that require improvement. Questionnaire surveying involves the use of surveys which are conducted to gather feedback from building occupants regarding their satisfaction levels and experiences with the building environment [35]. The survey questions cover various aspects of building design, IEQ, and other factors that may influence occupant satisfaction. The survey results provide valuable insights into occupants' perspectives and help identify areas where improvements can be made to enhance user satisfaction and comfort.

3.2 Case study

For the actualization of the research aim and objectives, it was necessary to collect relevant data from higher education institutions, primarily from the respective buildings associated with them. To achieve this, various factors such as the ease of collecting data, permissions required etc. were at the forefront of the selection process. After careful appraisal of the factors, two universities were identified and selected to serve as the study settings for this research with one being the main setting and the other being the complimentary setting. The study focused on assessing the state of IEQ in different types of rooms in the two universities (To satisfy ethical purposes, considerations, and requirements, the Universities will be addressed in this article as University A and University B). The different types of rooms, their respective capacity as well as their ventilation type (which is a key factor that studies have shown to affect IEQ) are represented in **Table 6**.

3.3 Data collection methods

As part of the POE process to assess the indoor environmental quality (IEQ) in the selected rooms at aforementioned universities, a comprehensive data collection approach was employed. These data collection methods involved physical measurements and questionnaire surveying with the collected data subjected to rigorous data analysis including the use of statistical methods, data visualization techniques, and qualitative analysis.

3.3.1 Physical measurements

Physical measurements involved undertaking a meticulous collection of data pertaining to key physical parameters that define the indoor environmental quality (IEQ) of the space under study. These include temperature (°C), relative humidity (RH %), CO_2 levels (ppm), airborne particulate matter (PM2.5—$\mu g/m^3$), total volatile organic compounds (TVOC—$\mu g/m^3$), formaldehyde (HCHO—$\mu g/m^3$), illuminance (lux), and background noise (dB).

To achieve the recording process, the researcher employed the use of certain physical measurement procedures after rigorous review of the literature. This includes the use of instrumentation which involves the use of appropriate instruments and sensors specific to each IEQ indicator. Datalogging techniques were also employed to continuously log the relevant data over a defined period. Field measurements were

University	Room	Capacity	Ventilation Type
A	Classroom (A)	56	Natural
	PC room (A)	24	Natural
	Lecture theater (A)	191	Mechanical
	Architectural studio (A)	30	Natural
B	Classroom (B)	41	Natural
	Lecture hall (B)	167	Mechanical

Table 6.
University data.

DOI: http://dx.doi.org/10.5772/intechopen.113332

conducted, including spot measurements at different times of the day, to capture variations in the IEQ indicators. Observation and documentation were integral to the data collection process, ensuring accurate and comprehensive records of the physical measurements.

By monitoring these IEQ indicators using suitable instruments and sensors, employing datalogging techniques, conducting field measurements, and documenting observations, a thorough assessment of the indoor environmental quality was achieved. This data serves as valuable information for evaluating the performance of the spaces and informing any necessary improvements or interventions.

3.3.2 Equipment used

The following highlighted equipment was used for the purpose of collecting physical IEQ data. The equipment was placed in a suitable location on the classroom walls at 1.5 m from the ground as that is best practice as seen in **Figure 3d**.

- Tinytag Datalogger (**Figure 3a**)—This is a compact, battery-operated instrument designed to record data over a set duration. It can be set up to take readings at consistent intervals, which could range from minutes to hours, based on the monitoring needs. The device saves the gathered data in its internal memory or storage. In the context of this study, the environmental factors measured included temperature, relative humidity, and carbon dioxide (CO_2).

- Temtop M2000C (**Figure 3b**)—This is a compact and portable air quality monitor which is powered by a rechargeable battery allowing for continuous monitoring. The device allows for real-time monitoring, datalogging and measures several environmental and air quality parameters including temperature, relative humidity, PM2.5, PM10, and CO_2.

- Temptop LKC 1000S+—This is a compact and portable air quality monitor which is powered by a rechargeable battery allowing for continuous monitoring. The device allows for real-time monitoring, datalogging and measures several environmental and air quality parameters including temperature, relative humidity, PM2.5, PM10, TVOC, and HCHO.

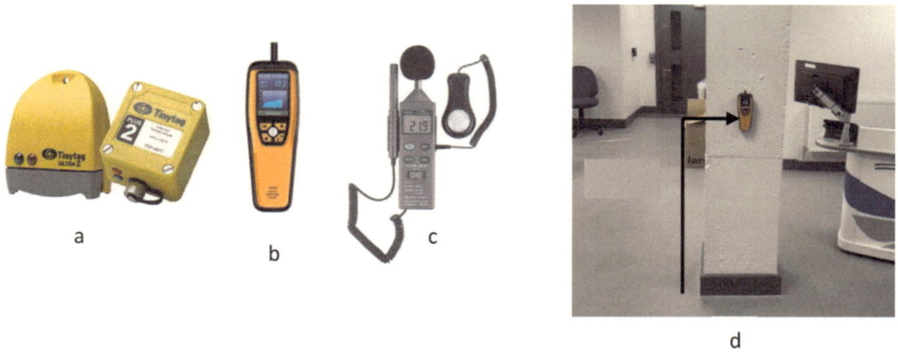

Figure 3.
Showing the equipment used and typical Temptop equipment placement in the classroom.

- Precision Gold N09AQ 4-in-1 environment meter (**Figure 3c**)—This is a versatile, handheld device designed to measure and monitor various environmental parameters. It is equipped with a clear LCD display screen which provides an easy-to-read measurement. The device typically includes built-in sensors for measuring temperature, humidity, light intensity, and sound level.

3.3.3 Surveys and questionnaires

Questionnaires are an effective and efficient method for gathering subjective feedback from occupants (students & staff) about their experiences and perceptions of indoor environmental quality. For this study, certain considerations and potential areas of inquiries were identified which included: overall satisfaction, thermal comfort, air quality, lighting conditions, and acoustic comfort.

4. Data collection outcomes

This section provides an overview of the data collection process and outcomes in this study. The data collection included two main types: quantitative data obtained through measurements and monitoring of various spaces in the two universities, and qualitative data gathered from questionnaire surveys. The objective of these data collection efforts was to assess the existing indoor environmental quality (IEQ) conditions in the higher education buildings. The measurements and surveys were conducted during the winter period, specifically between January and March 2023, in alignment with the seasonal conditions in the UK. By collecting these data, the study aimed to gain insights into the IEQ conditions and inform potential areas for improvement in the studied buildings.

4.1 IEQ results gauged against published standards and guidelines

To highlight major findings from the IEQ data, it is important to gauge it against the published standards and guidelines such as the CIBSE and BB101. This process is conducted under the four IEQ factors whereby the respective environment parameters such as temperature, CO_2, and so on are gauged against their respective standards.

4.1.1 Thermal and humidity data

Figure 4 illustrates the temperature data collected from various rooms at University A, along with the recommended temperature ranges (20–25°C) provided by the BB101 guidelines. The graph clearly shows that, except for the lecture theater, all other rooms exceeded the upper limit of 25°C, indicating a deviation from the recommended range. Of particular concern is the architectural studio, where temperatures consistently reached or exceeded 25°C for an extended period, with a peak of 29°C. The lecture theater and the PC room also experienced prolonged periods with temperatures around 25°C.

Figure 5 illustrates the relative humidity data obtained from various rooms at University B. The graph also includes the recommended optimum levels of relative humidity as specified by the BB101 guidelines. It is evident from the graph that the two rooms exhibited contrasting data. The classroom largely maintained relative

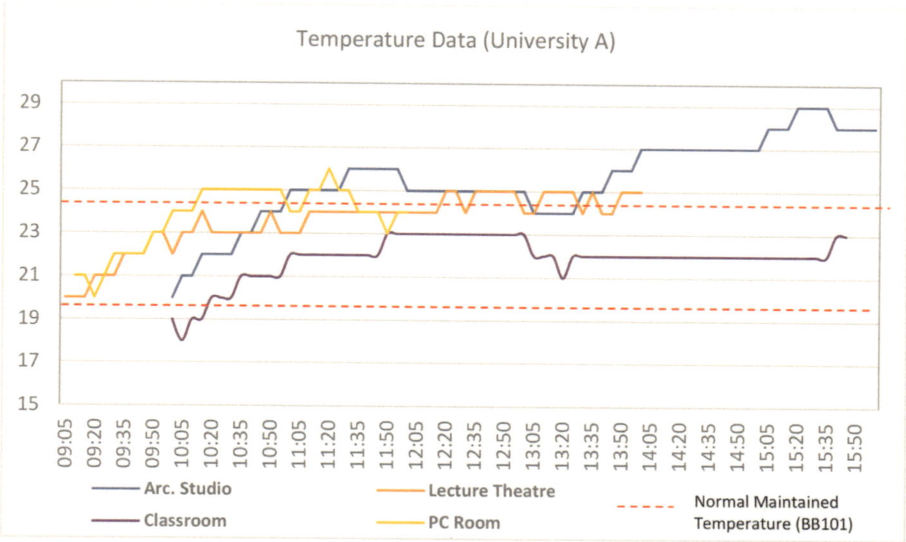

Figure 4.
Plotted temperature data for University A.

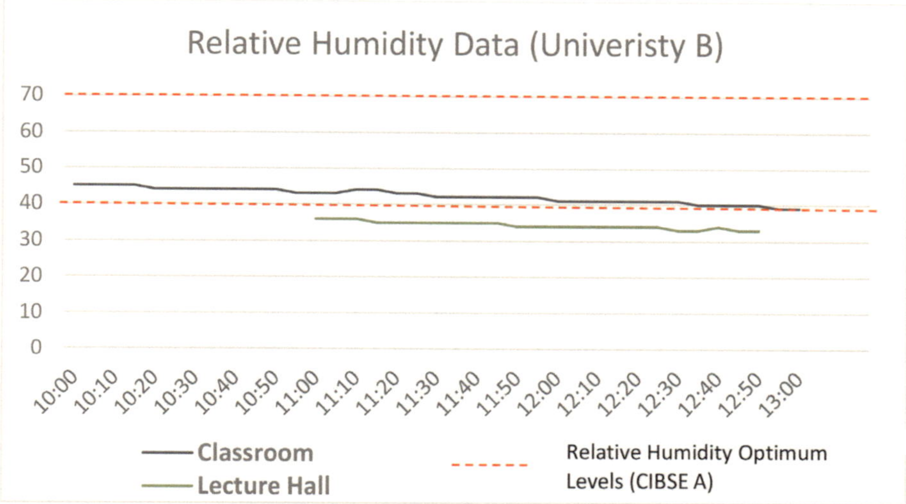

Figure 5.
Plotted relative humidity data for University B.

humidity within the recommended range of 40–70%, whilst the lecture hall recorded levels below the optimum range.

4.1.2 Indoor air quality data

Figure 6 displays a sample CO_2 data retrieved from all the selected rooms from University A and B including the published CO_2 limit of 1500 ppm. It can be observed from the graph that the architectural studio and the PC rooms especially recorded CO_2 levels reaching and exceeding the recommended limit.

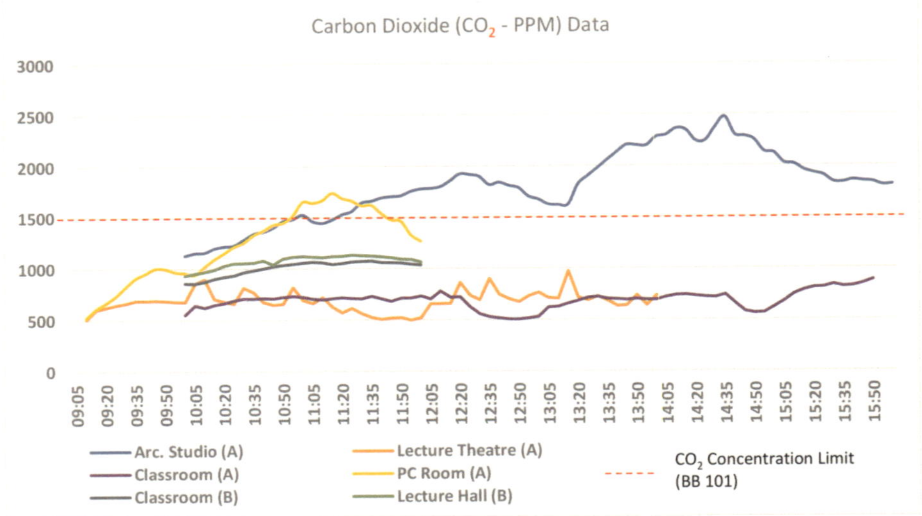

Figure 6.
Plotted CO_2 data for all rooms in University A & B.

The typical PM2.5 data for all the rooms in both universities is presented in **Figure 7**. This included the 24-h mean PM2.5 limit of 10 μg/m^3 as published by the UK Government. It can be observed that the PC room, architectural studio, and classroom, all of which are in University A, exceeded the mean limit.

The typical TVOC data for all the rooms in both universities is presented and plotted in **Figure 8**. This included the 8-h TVOC limit of 300 μg/m^3 as published by the BB101. It can be observed that all the rooms fell below the published limit.

The typical HCHO data for all the rooms in both universities is presented and plotted in **Figure 9**. This included the 30-min HCHO limit of 300 μg/m^3 as published by the UK Government. It can be observed that only the PC room recorded HCHO levels which exceeded the limit which even exceeded 30 min.

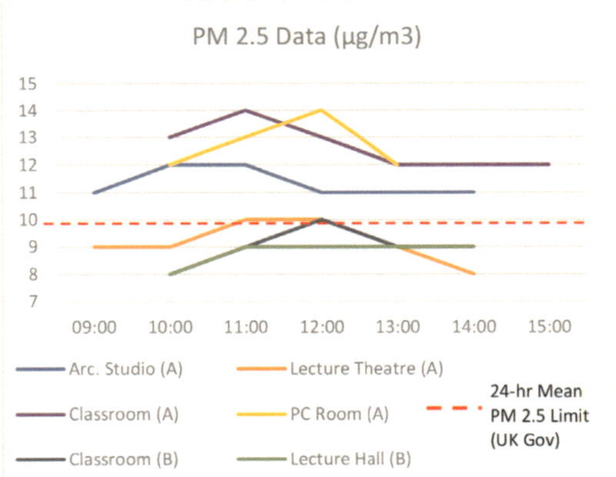

Figure 7.
Typical PM2.5 data for all rooms in both universities.

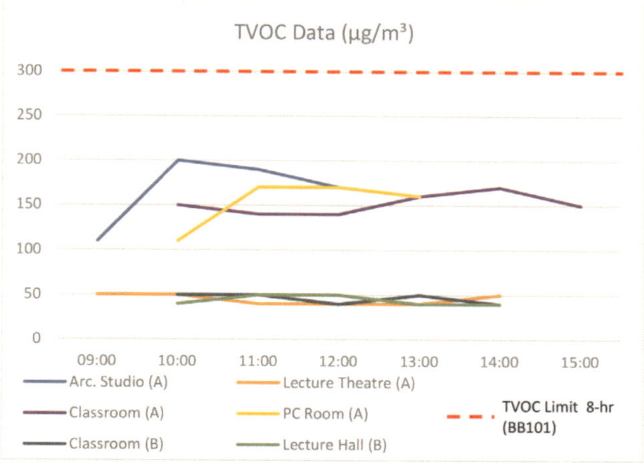

Figure 8.
Typical TVOC data for all rooms in both universities.

4.1.3 Visual comfort

The illuminance data collected from the classrooms showed a substantial compliance with the established guideline of 500 lux as shown in **Figure 10**. Most of the readings fell within this limit, indicating that the lighting conditions in the classrooms were generally appropriate for educational activities. However, there were instances where the illuminance levels exceeded the recommended limit, especially in University A. Upon further investigation, it emerged that these instances were primarily due to the presence of natural light, particularly when the curtains were open. In contrast, the classroom in University B fell short of the illuminance standard substantially, likely due to the type of lighting fixtures and the presence of light-limiting blinds which limited the entry of natural light. Whilst natural light can enhance the learning environment, it can evidently also lead to higher illuminance levels which can exceed a 1000 lux [36]. Therefore, it is important to manage the balance between natural and artificial light to maintain optimal lighting conditions in the classrooms.

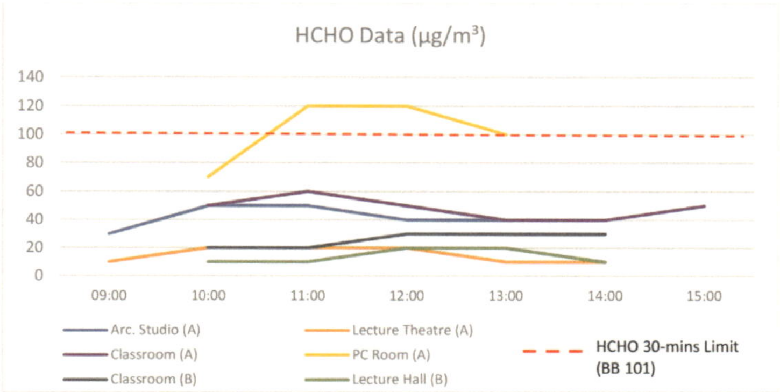

Figure 9.
Typical HCHO data for all rooms in both universities.

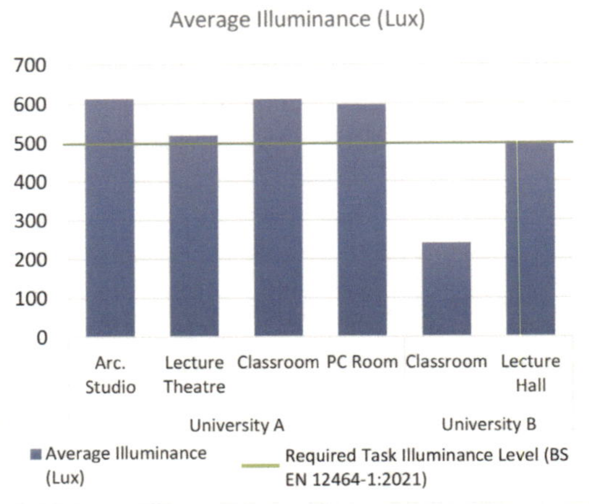

Figure 10.
Average illuminance for all rooms.

4.1.4 Acoustic comfort

The noise data, as shown in **Figure 11**, collected for both unoccupied and occupied rooms showed a general adherence to the expected noise levels. In unoccupied rooms, the noise levels substantially fell within the 30–35 dB range. However, slight variations were observed in situations where external factors, such as open windows, allowed noise from passing traffic to infiltrate the rooms, thus reflecting in the data. In occupied rooms, the noise levels were typically around 45 dB. This limit was occasionally surpassed during data collection periods when a lecturer was speaking, contributing to an increase in the ambient noise level. It is important to note that these occasional exceedances of the noise limit could be attributed to specific activities within the room rather than a consistent issue with noise control.

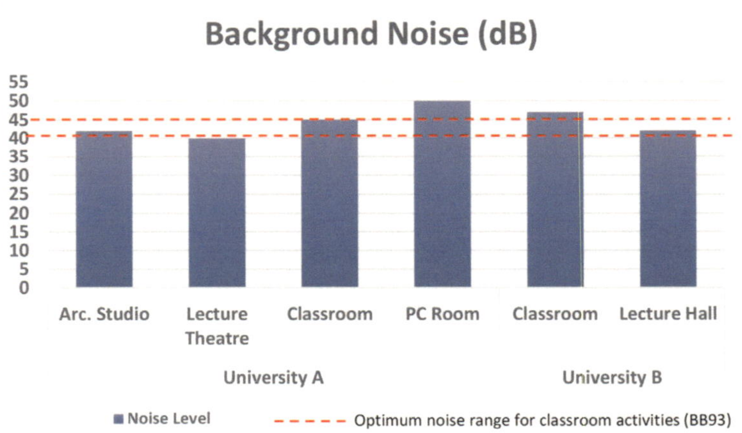

Figure 11.
Background noise levels for all rooms.

4.2 Survey feedback

The questionnaire survey was designed to gather occupants' satisfaction with the indoor environmental quality, focusing on four key aspects: temperature and humidity, air quality, lighting, and noise levels. A total of 41 feedback responses were collected from the spaces where physical monitoring and measurements took place. The surveys were typically conducted at the conclusion of teaching sessions. The survey questions and feedback provided valuable insights into occupants' perceptions and opinions regarding the indoor environment, offering the following perceptions and hypotheses:

4.2.1 Temperature perception

Amongst the questions asked are questions regarding the perception of temperature in classrooms by the students. The analysis, as represented in **Figure 12**, showed that majority of respondents described the room temperature as neutral, indicating overall comfort. However, a notable percentage felt it was either warm or cool, suggesting some variability in thermal conditions in the classrooms. Furthermore, emphasis was made for the heating season with most of the respondents having a moderate temperature perception, indicating effective heating system performance. However, a smaller percentage felt it was cool/cold, indicating a need for improvement during colder periods (**Figure 13**).

4.2.2 Air quality perception

Looking at air quality outlook in the classrooms was also important to the overall objectives of the research. Majority of respondents, as shown in **Figure 13**, perceived the air quality as good, indicating that it met their expectations contributing to a comfortable learning environment. However, a significant number reported bad air quality, suggesting the presence of issues that should be addressed to ensure optimal indoor air quality. Additionally, a significant percentage of respondents indicated their willingness to take action if they were informed of poor air quality, such as opening or closing windows. This highlights the importance of providing students with

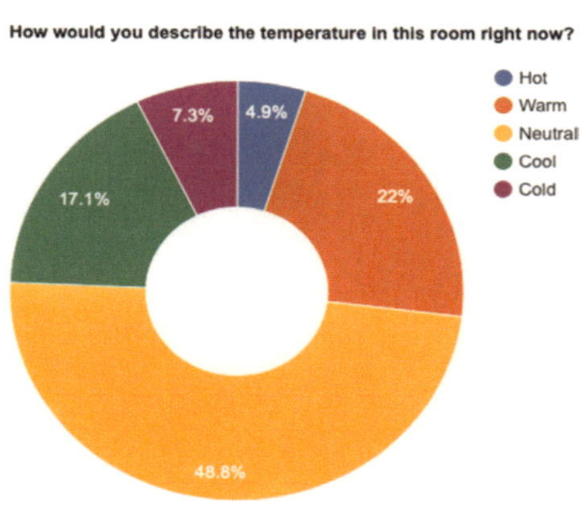

Figure 12.
Temperature perception feedback.

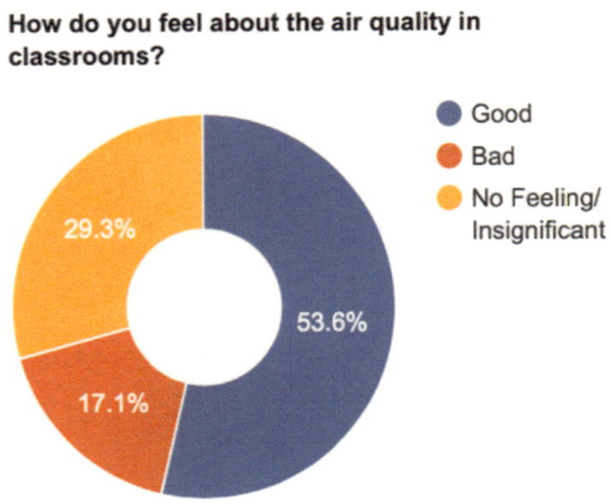

Figure 13.
Air quality perception feedback.

information and control over their environment, empowering them to improve their immediate surroundings.

4.2.3 Light level perception

Amongst the IEQ indicators is light which was also considered for this research. On this, most respondents perceived the light level in the classrooms as okay as shown in **Figure 14**, indicating that it was generally satisfactory for reading and visual tasks. However, a small percentage of respondents found the light level either too bright or dim, suggesting the need for adjustments or enhancements in lighting design.

4.2.4 Noise level perception

Another important IEQ indicator is the noise level which was also considered in the survey process. Noise levels from outside were generally deemed acceptable as shown in **Figure 15**, but occasional severe noise was reported, which could disrupt concentration. Notably, a significant percentage acknowledged that noise in the classroom does affect their concentration, emphasizing the importance of noise control measures to provide for a conducive learning environment.

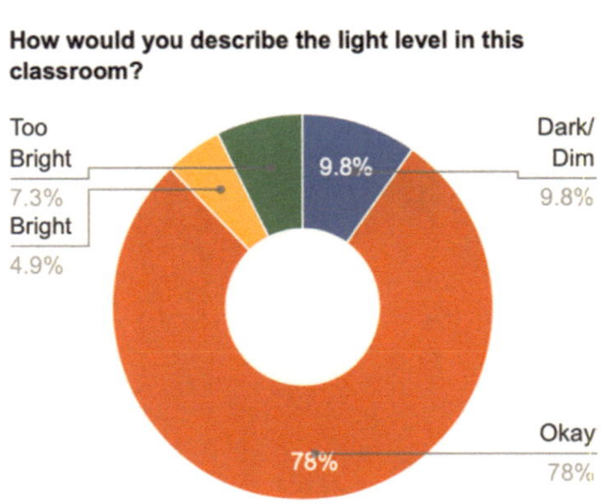

Figure 14.
Light perception feedback.

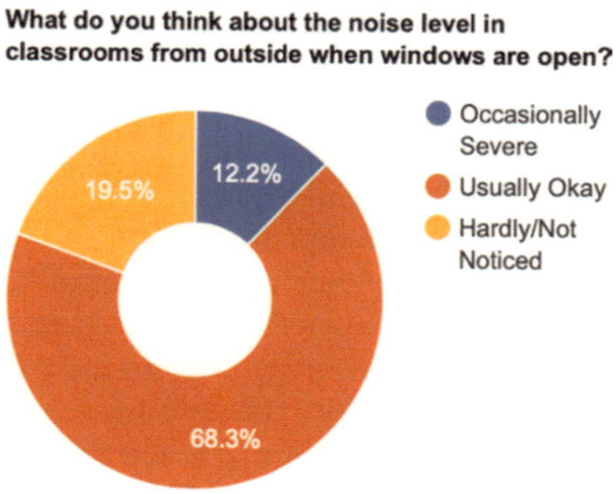

Figure 15.
Noise perception feedback.

4.2.5 Miscellaneous

Finally, students recognized the overall impact of IEQ on concentration, with temperature ranking as the most influential factor, followed by air quality, noise, and lighting. These findings provide valuable guidance for improving IEQ in higher education buildings, aiming to create comfortable, healthy, and conducive learning environments that optimize students' well-being and academic performance. (as well as IEQ added knowledge such as weighting of IEQ factors).

5. Analysis and discussion

5.1 Thermal and humidity

The summarized temperature data presented in **Table 7** reveals that a significant number of the rooms subjected to IEQ physical measurements (5 out of 6) exceeded the published upper limit of 25°C by the BB101. In University A, where the classrooms, PC room, and architectural studio are naturally ventilated, the maximum temperatures recorded during the winter months were relatively high. The classroom and PC room reached maximum temperatures of 28°C for 4 h and 5 min, and 3 h respectively, whilst the Architectural Studio recorded the highest maximum

University	Room	Min–Max temperature recorded °C	Duration below 20° C (h/min)	Duration Exceeded 25°C (h/min)
A	Classroom	19–26°C	0 h 0 min	4 h 5 min
	PC room	18–28°C	0 h 20 min	3 h 0 min
	Lecture theater	17–26°C	1 h 0 min	1 h 45 min
	Architectural studio	20–29°C	0 h 0 min	3 h 10 min
B	Classroom	20–23°C	0 h 0 min	0 h 0 min
	Lecture hall	21–26°C	0 h 0 min	0 h 15 min

Table 7.
Temperature data.

temperature of 29°C for 3 h and 10 min. In University B, the classroom recorded temperatures within the recommended range of 20–25°C, with a minimum and maximum temperature of 20–23°C. This suggests that the heating systems or insulation in the classroom were able to maintain a suitable temperature level during the winter months. However, the lecture hall, despite being mechanically ventilated, exceeded the upper limit, reaching a maximum temperature of 26°C. It is worth noting that it is located in the basement level of the university which combined with the heating systems in place, might have contributed to the temperature rise.

These findings are particularly noteworthy considering that the data was recorded during the winter period, when the outdoor temperatures ranged between 0 and 10°C. These results suggest that there may be challenges in maintaining optimal thermal comfort in the classrooms during colder months, potentially leading to discomfort for students and faculty. This observation suggests a potential issue with temperature control in the classrooms identified by the researcher, particularly during the colder months. One possible explanation for this discrepancy is the placement of temperature sensors predominantly in the corridors rather than within the classrooms themselves. As a result, the heating system may be working harder to maintain the desired temperature, leading to higher temperatures inside the classrooms. Additionally, this highlights the need for further investigation into the heating systems and insulation in these buildings to ensure a conducive learning environment, especially during extreme weather conditions.

For relative humidity, overall, the levels were found to be largely within the recommended standard range of 40–70% as indicated in **Table 8**. However, upon closer examination of the daily data, it became evident that the relative humidity levels tended to be on the lower side of the recommended range specifically in the 40% range. This suggests that the indoor environments in these rooms were relatively dry, approaching the lower end of the ideal humidity range.

In contrast, the lecture hall in University B displayed a lower relative humidity range, with a minimum and maximum of 33–36% respectively. This lower range might be attributed to the fact that the hall is mechanically ventilated which could impact humidity control of the space. But more importantly is the fact that it is located in the basement level of the university. Basements tend to have different environmental conditions compared to above-ground spaces, including potentially higher moisture levels. Considering the basement location and the potential challenges in humidity control, it is expected that the lecture hall in University B experienced lower relative humidity levels. If the recorded relative humidity falls outside the

University	Room	Relative humidity (%)	
		Minimum	Maximum
A	Classroom	41	46
	PC room	42	49
	Lecture theater	42	49
	Architectural studio	46	49
B	Classroom	39	45
	lecture hall	33	36

Table 8.
Relative humidity data showing the minimum and maximum recorded levels.

recommended range, it may be necessary to investigate further and implement measures to optimize humidity control, such as adjusting ventilation systems or introducing additional humidity management strategies, to ensure a comfortable and healthy indoor environment for occupants. Additionally, **Figure 5** visually demonstrates that there were instances where the relative humidity did not meet the lower limit of the standard in certain classrooms. The researcher observed that the windows for rooms with natural ventilation were open on many occasions which could influence the relative humidity levels recoded. However, this issue still suggests a potential challenge in maintaining optimal humidity levels within the classrooms which highlights the need for further investigation into the causes of these deviations.

5.2 Indoor air quality

5.2.1 Carbon dioxide (CO_2)

Table 9 provides an overview of the maximum CO_2 levels recorded in various rooms of two universities, along with the number of times the CO_2 concentration exceeded 1500 ppm and the duration of such exceedances. When analyzing the data from both universities, it was observed that the rooms that exceeded the limit occurred in university A specifically, the classroom, PC room, and architectural studio. In University B, both the classroom and lecture hall maintained CO_2 levels below the threshold, implying better ventilation and lower occupancy in these rooms from the observations.

Upon further investigation, it became evident that the occurrence of high CO_2 levels primarily affected classrooms that relied on natural ventilation. Specifically, the Architectural studio, with the plan as shown in **Figure 16**, recorded the highest CO_2 level of 2360 ppm. This room is naturally ventilated and has two small windows, but only one window can be opened for airflow. Additionally, it is worth noting that the usable open window has a relatively small area of 0.93 m^2, which may not be sufficient for the room size of 85 m^2. This finding can be attributed to several factors related to the studio's ventilation and occupancy conditions. Firstly, the limited use of one window hindered the proper airflow and exchange of fresh air, resulting in a build-up of CO_2. Inadequate ventilation promotes stagnant air and the accumulation of pollutants. Secondly, the studio was frequently overcrowded, with a large number of occupants present. As people exhale, they release CO_2, and in a crowded space, CO_2 concentration can rise rapidly. The combination of limited ventilation and high

University	Room	Maximum CO_2 recorded (ppm)	Duration exceeded 1500 ppm (h/min)
A	Classroom	1617	1 h 45 min
	PC room	1738	1 h 0 min
	Lecture theater	970	0 h 0 min
	Architectural studio	2360	8 h 55 min
B	Classroom	1072	0 h 0 min
	Lecture hall	1133	0 h 0 min

Table 9.
CO_2 data.

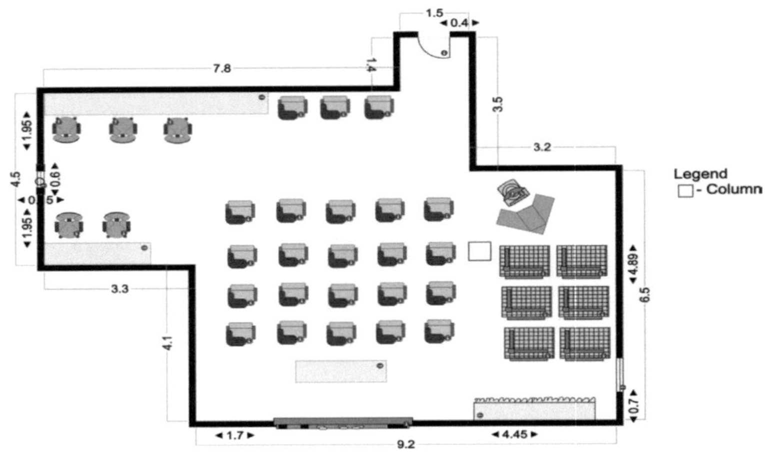

Figure 16.
Architectural studio plan.

occupancy levels likely contributed to the observed elevated CO_2 levels. These findings underscore the importance of optimizing ventilation strategies and ensuring an adequate supply of fresh air in crowded spaces. By doing so, it is possible to maintain healthy indoor air quality and minimize CO_2 build-up in classrooms relying on natural ventilation (**Figure 16**).

5.2.2 PM2.5, TVOC, and HCHO

The PM2.5 data showed that in some classrooms, on certain days, the hourly PM 2.5 readings exceeded the daily limit of 10 $\mu g/m^3$ (average) as represented in **Figure 7**. This primarily points to PM2.5 being a slight concern considering the case study settings are located in an urban area with high traffic which affects the quality of outdoor air. This in return affects the indoor PM2.5 readings with other possible factors being classroom conditions or student activities.

With regard to TVOC and HCHO, the analysis of these data sets revealed that the levels largely adhered to the established limit of 300 $\mu g/m^3$ or 100 $\mu g/m^3$ respectively. This standard was exceeded only on one or two occasions, suggesting a high level of compliance with the recommended guidelines. The few instances where especially the HCHO levels surpassed the limit could potentially be attributed to increased student activity, which might have led to a temporary spike in emissions, or possibly equipment error. However, these instances were exceptions rather than the norm, indicating that the indoor environment was generally within the acceptable range for TVOC/HCHO levels.

5.3 Room IEQ correlations

In the process of data analysis, correlations were carried out with the aim of examining various variables against each other in the bid to ascertain various findings align with those from prior research. The relationship between occupant density and temperature and CO_2 levels was examined for rooms in University A with **Table 10** presenting the results and data. The results indicated a weak correlation between occupant density and temperature, suggesting that the number of occupants in a room may not significantly impact the temperature. However, a strong positive linear

		PC room		Classroom		Lecture theater		Architectural studio	
		Occ. density	Count	Occ. density	Count	Occ. density	Count	Occ. density	Count
CO_2 concentration	Pearson correlation	0.7465	0.7632	0 7469	0.7375	0.5032	0.5081	0.7233	0.681
	Sig. (2 tailed)	0.0001	0.0000	0.0000	0.0000	0.0000	0.0000	0.0001	0.0001
		Correlation is significant at p < .05 and at p < .10 for both occupant count and density		Correlation is significant at p < .05 and at p < .10 for both occupant count and density		Correlation is significant at p < .05 and at p < .10 for both occupant count and density		Correlation is significant at p < .05 and at p < .10 for both occupant count and density	

Table 10.
CO_2 and occupant density correlation data.

correlation was observed between occupant density and CO_2 levels. This implies that as the number of occupants increases, the CO_2 levels in the room also rise, which is consistent with the findings from previous literature (**Table 10**) [37].

5.4 Room comparison: mechanical and natural ventilation

To further scrutinize the IEQ data, it was pertinent to look at relationships and make comparisons to distinctive differences between classrooms such as looking at the data outlook with regard to mechanically and naturally ventilated classes. After undergoing various comparisons between the classrooms and the respective IEQ indicators, the researcher realized that the CO_2 data exhibited the most glaring differences when the two types of classrooms were compared. The analysis showed that CO_2 levels in mechanically ventilated rooms consistently remained below the recommended maximum limit of 1500 ppm, while naturally ventilated rooms occasionally exceeded this limit as shown in **Figure 17**. This discrepancy can be attributed to the active control and circulation of air in mechanically ventilated rooms, as opposed to the passive air movement in naturally ventilated rooms, which can be less effective, especially in high occupancy spaces or rooms with limited airflow (**Figure 17**).

5.5 Combined IEQ and survey findings

The analysis of both the IEQ data and survey feedback presented various results and findings. Incidentally, analyzing both the objective IEQ data and the subjective survey responses, meaningful connections, and insights were highlighted as such:

1. *Temperature*: A notable percentage of respondents felt their classrooms were balanced, but a fair amount (4.9%) perceived the temperature to be hot. This aligns with the IEQ data, which indicates that the temperature in classrooms reached or exceeded the recommended maximum limit of 25°C. The discomfort expressed by students during these situations, leading to the removal of top layers such as coats and the opening of windows, further supports this alignment. Research has shown that the optimal temperature for a learning environment is

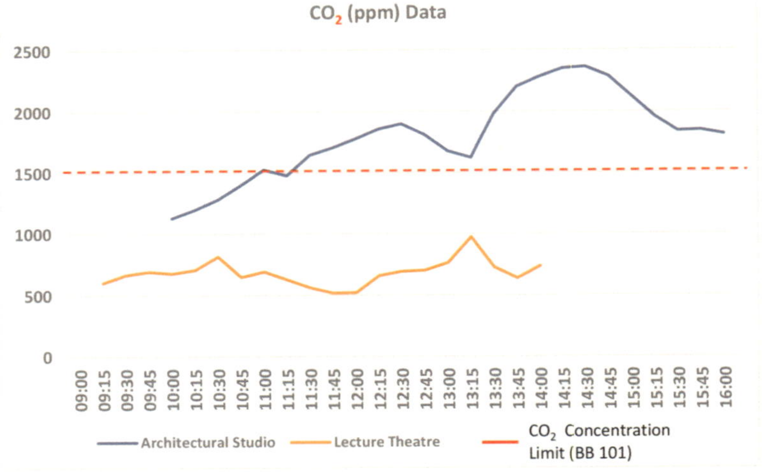

Figure 17.
CO_2 outlook for a typical mechanically and naturally ventilated classroom.

between 18°C and 25°C, as temperatures outside of this range can impact students' focus and learning abilities [38].

2. *Air quality*: Whilst a sizeable percentage of respondents felt the air quality was okay, a notable percentage (17.1%) expressed it as bad. This strongly aligns with the IEQ data, particularly the CO_2 levels, which recorded high levels exceeding the maximum limit of 1500 ppm, especially in naturally ventilated classrooms. The high CO_2 levels recorded in the IEQ data correspond to the dissatisfaction expressed by the respondents regarding air quality with elevated CO_2 levels being able to impact cognitive function and contribute to feelings of stuffiness and discomfort [3].

3. *Lighting*: The survey responses and IEQ data show some alignment in terms of lighting. Whilst a majority of respondents felt the lighting was okay, a notable percentage found it to be bright or too bright. This aligns with the IEQ data indicating that the illuminance in classrooms exceeded the recommended limit of 500 lux. The perception of bright lighting by some respondents is consistent with the measured illuminance levels.

4. *Overall IEQ*: The finding that most respondents agreed that the overall indoor environmental quality affected their concentration in the classroom is a strong indicator of student awareness regarding IEQ. This alignment also corresponds to the IEQ data, which identified different indicators, including temperature and CO_2 levels, exceeding their recommended limits. The impact of IEQ on students' well-being, comfort, and concentration underscores the importance of addressing and improving indoor environmental conditions [39].

5.6 Research limitations and challenges faced

A few limitations or challenges were experienced by the researcher during the POE data collection process. Firstly, the research encountered technical gaps in equipment, which had a direct impact on the accuracy and precision of measurements. The quality and capabilities of the equipment used ultimately influenced the reliability of the data collected. Limited resources and funding also posed challenges in accessing specialized or advanced measurement tools, further compromising the evaluation process. This limitation particularly affected the assessment of specific building performance indicators, such as indoor air quality, which required expensive and less readily available equipment. Consequently, the depth and comprehensiveness of the evaluation were restricted. Moreover, the limited resources also had implications for the scope and scale of the evaluation, resulting in a narrower focus and reduced sample size. Additionally, establishing standardized metrics and benchmarks for physical measurements in higher education buildings proved to be a challenging task. The lack of uniformity in measurement protocols, criteria, published standards, and guidelines hindered the ability to compare the performance of different buildings or assess their performance against established benchmarks. For example, after thorough scrutiny of the available published standards and guidelines, the researcher found that many of them focused on primary and secondary schools. This limitation made it difficult to draw meaningful conclusions and accurately evaluate the buildings' IEQ performance. The absence of a standardized framework limited the researcher's ability to make comprehensive and reliable comparisons, thus slightly hampering the outcomes in the field of POE in higher education buildings.

6. Conclusions

The post-occupancy evaluation (POE) of higher education buildings provided invaluable insights into the indoor environmental quality (IEQ) within these educational spaces. This study identified several critical IEQ issues associated with the factors that predominantly affect the state of IEQ.

Thermal discomfort emerged as a significant issue, with temperature and relative humidity often not meeting standard limits. This was particularly pronounced in rooms without heating/cooling control and spaces with either natural or mechanical ventilation. To address this, it is recommended that temperature control systems be installed and ensuring adequate ventilation in all rooms. Air quality represented another major concern, with elevated levels of CO_2, volatile organic compounds (TVOC), formaldehyde (HCHO), particulate matter (PM 2.5) exceeding established standard limits. Factors such as inadequate and insufficient ventilation as well as high traffic within the HE locations contributed to the problems. Recommendations include enhancing ventilation systems and possibly introducing air purifiers to mitigate these concerns. Noise pollution was identified as an issue, particularly in natural ventilated rooms and classrooms. To alleviate this, implementing soundproofing measures and reinforcing classroom behavior guidelines may prove effective. Inadequate lighting was occasionally observed, negatively affecting visibility and potentially causing eye strain and reduced productivity. This could be addressed by re-evaluating and improving lighting placement and levels in classrooms. Ergonomic issues were also identified, with room layouts, furniture, and equipment sometimes failing to meet their intended purpose. It is recommended to comprehensively review and adjust these elements to ensure they are suitable for their intended use. Additionally, this study identified maintenance issues and a lack of access to green spaces as areas of concern. To address this, implementation of regular maintenance schedules to ensure that all building systems are functioning optimally is recommended.

Furthermore, incorporating green spaces or elements of nature into building designs could enhance IEQ and the overall well-being of the students and staff alike.

These findings underscore the critical importance of IEQ in higher education buildings and the urgent need for effective strategies to enhance it. The feedback from building occupants reinforced the significance of these IEQ challenges, with a considerable percentage expressing discomfort with temperature and air quality, aligning with the recorded IEQ data. Most respondents acknowledged that IEQ significantly influenced their concentration in the classroom, indicating a heightened awareness of IEQ amongst students. Furthermore, this study and its findings highlight the value of POE methods in assessing IEQ and providing actionable insights for improving the indoor environment. Moving forward, it is recommended that future research continues to use POE methods to assess IEQ in higher education buildings. Doing so will not only help improve the design and operation of these buildings but also contribute to the growing body of knowledge on the relationship between IEQ and occupant satisfaction.

Author details

Mukhtar Maigari, Changfeng Fu*, Jie Deng and Ali Bahadori-Jahromi
The School of Computing and Engineering, The University of West London, London, UK

*Address all correspondence to: charlie.fu@uwl.ac.uk

References

[1] Kamaruzzaman SN, Egbu C, Mahyuddin N, Ahmad Zawawi EM, Chua SJL, Azmi NF. The impact of IEQ on occupants' satisfaction in Malaysian buildings. Indoor and Built Environment. 2017;**27**:715-725. Available from: https://www.semanticscholar.org/paper/The-impact-of-IEQ-on-occupants%E2%80%99-satisfaction-in-Kamaruzzaman-Egbu/9fc8fce2e3d7e86c3b92c103aea78fdae0beffb0 [Accessed: June 12, 2023]

[2] Awang NA, Mahyuddin N, Kamaruzzaman SN. Indoor environmental quality assessment and user's perception in Meru Secondary School (Smk Meru). Journal of Building Performance. 2015;**6**(1):105-115. Available from: https://spaj.ukm.my/jsb/index.php/jbp/article/view/160 [Accessed: June 12, 2023]

[3] Korsavi S, Montazami A, Mumovic D. Indoor air quality (IAQ) in naturally-ventilated primary schools in the UK: Occupant-related factors. Building and Environment. 2020;**180**: 106992. DOI: 10.1016/j.buildenv.2020.106992

[4] Leaman A. Post-occupancy Evaluation, Usable Buildings. 2003. Available from: https://www.usablebuildings.co.uk/UsableBuildings/Unprotected/AdrianLeamanPost-OccupancyEvaluation.pdf [Accessed: May 15, 2023]

[5] Kamaruzzaman SN, Egbu C, Mahyuddin N, Ahmad Zawawi EM, Chua SJL, Azmi NF. Effect of Indoor Environmental Quality (IEQ) to the Human Occupation Health and Performance in Buildings. University of East London Repository; 2019. Available from: https://core.ac.uk/download/pdf/228124676.pdf [Accessed: October 24, 2023]

[6] NIOSH. Indoor Environmental Quality. The National Institute for Occupational Safety and Health (NIOSH); 2022. Available from: https://www.cdc.gov/niosh/topics/indoorenv/default.html#print [Accessed: May 03, 2023]

[7] Medium. Indoor Environmental Quality Parameter. Medium; 2020. Available from: https://medium.com/@ieqandwellbeing/indoor-environmental-quality-parameter-56aab5cffd8 [Accessed: April 29, 2023]

[8] Abdulaali HS et al. Impact of poor indoor environmental quality (IEQ) to inhabitants' health, wellbeing and satisfaction. International Journal of Advanced Science and Technology. 2020;**29**(3):1284-1296. Available from: https://www.researchgate.net/publication/340278976_Impact_of_poor_Indoor_Environmental_Quality_IEQ_to_Inhabitants'_Health_Wellbeing_and_Satisfaction

[9] Haverinen-Shaughness U et al. Clima - WellBeing Indoors. In: Indoor Environmental Quality in Schools in Relation to Academic Performance of Students: Observations of Potential Contributors to Poor IEQ. Helsinki: FINVAC; 2007. pp. 403-407

[10] ASHRAE. ANSI/ASHRAE Standard 55-2010. ASHRAE-RP; 2013. Available from: http://www.ashraerp.com/files/ASHRAEStandard55-2013.pdf [Accessed: May 16, 2023]

[11] ISO. International ISO Standard 7730, Standards. 2010. Available from: https://cdn.standards.iteh.ai/samples/39155/9632a0563ca742209edb45856ff69296/ISO-7730-2005.pdf [Accessed: May 16, 2023]

[12] Centnerová LH. Rehva Journal 02/2018—On the History of Indoor Environment and It's Relation to Health and Wellbeing. REHVA; 2018. Available from: https://www.rehva.eu/rehva-journal/chapter/on-the-history-of-indoor-environment-and-its-relation-to-health-and-wellbeing [Accessed: May 16, 2023]

[13] Ali SM. Measured and perceived conditions of indoor environmental qualities (IEQ) of university learning environments in semi-arid tropics: A field study in Kano-Nigeria [Doctoral thesis]. Portsmouth: University of Portsmouth; 2018

[14] Dorizas P, De Groote M, Volt J. The Inner Value of a Building. Copenhagen: Buildings Performance Institute Europe (BPIE); 2018. pp. 8-33

[15] Mannan M, Al-Ghamdi SG. Indoor air quality in buildings: A comprehensive review on the factors influencing air pollution in residential and commercial structure. International Journal of Environmental Research and Public Health. 2021;**18**(6):3276. DOI: 10.3390/ijerph18063276

[16] ASHRAE. Ventilation for Acceptable Indoor Air Quality, Upgreengrade. 2016. Available from: https://upgreengrade.ir/admin_panel/assets/images/books/25223276727.pdf [Accessed: May 17, 2023]

[17] Katiyar V, Khare M. Indoor Air Quality an Emerging Environmental Challenge. Research Gate; 2008. Available from: https://www.researchgate.net/profile/Vinita-Katiyar/publication/236605730_Indoor_Air_Quality-An_Emerging_Environmental_Challenge/links/00463518379af9fd71000000/Indoor-Air-Quality-An-Emerging-Environmental-Challenge.pdf [Accessed: May 17, 2023]

[18] Ralegaonkar RV, Sakhare VV. Development of multi-parametric functional index model for evaluating the indoor comfort in built environment. Indoor and Built Environment. 2013;**23**(4):615-621. DOI: 10.1177/1420326x13480515

[19] Abdul Mujeebu M. Chapter Introductory Chapter: Indoor Environmental Quality. London, UK: IntechOpen. Available from: https://cdn.intechopen.com/pdfs/65121.pdf; 2019 [Accessed: May 17, 2023]

[20] Vardaxis N, Bard D, Persson Waye K. Review of acoustic comfort evaluation in dwellings—Part I: Associations of acoustic field data to subjective responses from building surveys. Building Acoustics. 2018;**25**(2):151-170. DOI: 10.1177/1351010x18762687

[21] Al-Khatatbeh B, Ma'bdeh, S. Improving visual comfort and energy efficiency in existing classrooms using passive daylighting techniques. Energy Procedia. 2017;**136**:102-108. DOI: 10.1016/j.egypro.2017.10.294

[22] Abdelatia B, Marenne C, Semidor C. Daylighting strategy for sustainable schools: Case study of prototype classrooms in Libya. Journal of Sustainable Development. 2010;**3**(3):60-67. DOI: 10.5539/jsd.v3n3p60

[23] EN-European Standard 12665. Light and Lighting - Basic Terms and Criteria for Specifying Lighting Requirements. 2018. Available from: https://cdn.standards.iteh.ai/samples/60790/f5bfb7345fc94916bfed08ca97f5474a/SIST-EN-12665-2018.pdf

[24] Fakhari M, Vahabi V, Fayaz R. A study on the factors simultaneously affecting visual comfort in classrooms: A structural equation modelling

approach. Energy and Buildings. 2021; **249**:111232. DOI: 10.1016/j.enbuild.2021.111232

[25] Hwang T, Jeong TK. Effects of indoor lighting on occupants' visual comfort and eye health in a green building. Indoor and Built Environment. 2010;**20**(1):75-90. DOI: 10.1177/1420326x10392017

[26] DiLouie C. Lighting Redesign for Existing Buildings, Perlego. River Publishers; 2011. Available from: https://www.perlego.com/book/2096016/lighting-redesign-for-existing-buildings-pdf [Accessed: 2022]

[27] Sulaiman M, Yusoff W, Pawi S, Kamarudin W. Indoor environmental quality (IEQ) of higher education institutions (HEIs): A user perception survey. Journal of Clean Energy Technologies. 2013;**1**(4):318-321. DOI: 10.7763/jocet.2013.v1.72

[28] Education Funding Agency. Baseline Designs for Schools: Guidance. GOV.UK; 2014. Available from: https://www.gov.uk/government/publications/baseline-designs-for-schools-guidance/baseline-designs-for-schools-guidance [Accessed: June 03, 2023]

[29] European Commission. Key players in European standardisation, Internal Market, Industry, Entrepreneurship and SMEs. 2022. Available from: https://single-market-economy.ec.europa.eu/single-market/european-standards/key-players-european-standardisation_en [Accessed: 16 August 2022]

[30] CENCENELEC. About CENELEC. CEN-CENELEC [Online]. 2021. Available from: https://www.cencenelec.eu/about-cenelec/ [Accessed: August 18, 2022]

[31] ASHRAE. ASHRAE's Mission and Vision, Mission and Vision. 2023. Available from: https://www.ashrae.org/about/mission-and-vision#:~:text=ASHRAE%3A%20Shaping%20Tomorrow's%20Built%20Environment,and%20sustainability%20within%20the%20industry [Accessed: October 24, 2023]

[32] Hadjri K, Crozier C. Post-occupancy evaluation: Purpose, benefits and barriers. Facilities. 2009;**27**(1/2):21-33. DOI: 10.1108/02632770910923063

[33] Fisk D. Sustainable development and post-occupancy evaluation. Building Research and Information. 2001;**29**(6):466-468. DOI: 10.1080/09613210110072665

[34] Guerra-Santin O, Tweed CA. In-use monitoring of buildings: An overview of data collection methods. Energy and Buildings. 2015;**93**:189-207. DOI: 10.1016/j.enbuild.2015.02.042

[35] Tookaloo A, Smith R. Post occupancy evaluation in higher education. Procedia Engineering. 2015; **118**:515-521. DOI: 10.1016/j.proeng.2015.08.470

[36] Kong Z, Jakubiec JA. Instantaneous lighting quality within higher educational classrooms in Singapore. Frontiers of Architectural Research. 2021;**10**(4):787-802. DOI: 10.1016/j.foar.2021.05.001

[37] Franco A, Leccese F. Measurement of CO_2 concentration for occupancy estimation in educational buildings with energy efficiency purposes. Journal of Building Engineering. 2020;**32**:101714. DOI: 10.1016/j.jobe.2020.101714

[38] Mendell MJ, Heath GA. Do indoor pollutants and thermal conditions in schools influence student performance? A critical review of the literature. Indoor Air. 2005;**15**(1):27-52. DOI: 10.1111/j.1600-0668.2004.00320.x

[39] Shan X, Melina AN, Yang E-H. Impact of indoor environmental quality on students' wellbeing and performance in educational building through life cycle costing perspective. Journal of Cleaner Production. 2018;**204**:298-309. DOI: 10.1016/j.jclepro.2018.09.002

Chapter 4

Priority of Mixed-Mode Ventilation during Epidemics: A Comprehensive Investigation

Ijaz Fazil Syed Ahmed Kabir, Mohan Kumar Gajendran and Eddie Yin-Kwee Ng

Abstract

This chapter provides a detailed analysis of the operation of mixed-mode ventilation during epidemics, concentrating on the pivotal role of indoor air quality (IAQ). It underlines the importance of ventilation in IAQ management, particularly for airborne infection control. However, our principal focus is mixed-mode ventilation, a combined approach of natural and mechanical methods, which we highlight as promising for IAQ management, airborne disease control, and also energy-saving solutions. Our examination includes multiple case studies for each diverse environment, such as educational buildings, hospitals, office buildings, and residential buildings, each evaluated through different methods, including computational fluid dynamics and experimental approaches. Our observations illustrate the significant role of efficient ventilation in improving IAQ, mitigating airborne infection risks, and enhancing occupant comfort, especially during epidemics.

Keywords: indoor air quality, ventilation strategies, epidemics, airborne infections, mixed mode ventilation, mechanical ventilation, ventilation standards, international standards, energy efficiency, ventilation optimization, hybrid ventilation, case studies

1. Introduction

1.1 Importance of indoor air quality (IAQ)

The importance of indoor environment quality for occupant health cannot be overstated. The quality of indoor air influences the health of those who work, reside, or use enclosed spaces in the same way that outdoor air quality does. Experts emphasize the critical significance of indoor air quality, sanitation, ventilation, and pollution control in the context of the Covid-19 pandemic [1].

The phrase "indoor air quality," abbreviated as "IAQ," refers to the state of the air within a building or other enclosed space, taking into account both the thermal comfort levels and the concentrations of various pollutants [2–5]. ANSI/ASHRAE 62.1-2022 [6] defines acceptable indoor air quality (IAQ) as "air in which there are no known contaminants at harmful concentrations, as determined by cognizant authorities, and with which a substantial majority (80% or more) of the people exposed do not express dissatisfaction."

Because we spend the majority of our time indoors, IAQ is crucial. Occupants may believe that their enclosed space is exceptionally clean, but do they realize that indoor air contaminants are virtually everywhere? Construction materials, paints, and coatings on walls and ceilings, gas ranges in kitchens, and even personal care products contain contaminants that can be hazardous in an indoor environment. Indoor air can be several times more polluted than outdoor air. Because of this, experts are concerned about IAQ in contemporary times.

The contaminants in the indoor air lead to the degradation of the IAQ. Radon, indoor aerosols, ozone (O_3), carbon dioxide (CO_2), carbon monoxide (CO), formaldehyde (HCHO), nitrogen dioxide (NO_2), sulfur dioxide (SO_2), total volatile organic compounds (TVOCs), particulate matter (PM2.5 and PM10), and asbestos are all examples of contaminants that may be found in indoor air [2, 3, 7]. Temperature, humidity, air movement, and the effectiveness of ventilation systems are other factors that impact IAQ [3]. Modern buildings are built to be airtight, which eliminates the need for natural ventilation. This is because modern buildings are controlled by heating, ventilation, and air conditioning (HVAC) systems, which recycle a large proportion of the air while replacing minor amounts of it with fresh air in order to maintain a consistent IAQ throughout every season [7, 8]. Imagine a confined space that has a high level of humidity and inadequate ventilation. It should come as no surprise that this space has a poor IAQ. When one person in this enclosed space has influenza, it is quite likely that other persons in the same setting will also get infected with the virus. When someone remains with an individual who suffers from a major infectious condition like severe acute respiratory syndrome (SARS) or COVID-19, the situation is made much worse, and the outcomes may be fatal.

Thus, IAQ is essential. The concentrations of indoor air contaminants are maintained at an acceptable level in an indoor environment with adequate IAQ. The health of the occupants can then be protected. In addition, the indoor temperature, humidity, and ventilation system can all contribute to comfort. Conversely, an environment with inadequate IAQ can result in a variety of problems. In addition to making occupants feel uncomfortable, inadequate IAQ has negative effects on health. One of the most significant building-related health issues is the sick building symptom (SBS), which is caused by inadequate IAQ. SBS side effects include migraines, vertigo, fatigue, shortness of breath, eye irritation, dry cough, and itchiness on the skin [3, 7]. Hence, it is necessary to maintain a good IAQ within a building.

1.2 Importance of ventilation strategies in improving IAQ and mitigating airborne infections

In 2019, a worldwide outbreak of a hitherto undiscovered virus halted almost all human endeavors. Damage from this pandemic has been catastrophic. The COVID-19 epidemic is, by any measure, one of the most major issues faced by mankind in the twenty-first century [9–11].

As the World Health Organization (WHO) designated COVID-19 to be a Public Health Emergency of International Concern (PHEIC) on January 30, 2020 [12, 13], very few people possibly could have predicted that it would infect about 691,748,366 individuals and killed 6,901,518 people all over the globe by July 20, 2023. In addition, the world economy was on the edge of entering a downturn on a scale never seen before when the COVID-19 epidemic broke out. COVID-19 is a coronavirus, also referred to as severe acute respiratory syndrome coronavirus 2 (SARS-CoV-2) [14]. The prevalence of this virus has actually eclipsed the total amount of infections caused by two different epidemics (severe acute respiratory syndrome (SARS) and Middle East respiratory syndrome (MERS)) in this century [11, 15]. Patients who are old or who have previous pulmonary diseases have a greater risk of passing away as a result of it [9, 16].

According to the contamination data, the epidemic spread swiftly across the globe. As a result, governments were compelled to implement preventative measures, such as lockdowns, self-isolation, social isolation, the use of facial masks and shields, and the recommendation to cleanse hands as often as feasible [10, 11].

Even though it was once believed that the virus spread by droplets and means of contact and that social distance might assist in disrupting the transmission chain, recent scientific research has essentially proven the possibility that the virus transmits through the air instead and survives for up to 3 hours in the air [9, 11, 17, 18].

Due to the overwhelming majority of infections happening indoors [19], recently published studies have suggested and assessed several strategies to lower or eliminate any virus level, including ventilation systems, high-efficiency filtration, ultraviolet irradiation, air ionization, chemical disinfection, nonthermal plasma, and filter-based air cleaners [20–22]. Researchers have demonstrated that ventilation strategies are among the most effective techniques for reducing the possibility of viral transmission [10, 11, 20, 23–28].

In recent times, the Chartered Institution of Building Services Engineers (CIBSE) released guidance on how ventilation can be used to mitigate airborne infections. It has been reported that “there is good evidence that demonstrates room occupants are more at risk of catching an illness in a poorly ventilated room than in a well-ventilated room” [11].

The World Health Organization has stressed ventilation’s importance in improving indoor air quality and health on multiple occasions [10, 20, 29–33].

Air recirculation by ventilation can be the main route for aerosol transmission, increasing infection risks [10]. The aerosol movement was almost neglected in designing ventilation systems [10]. In addition, an inadequate rate of ventilation and ineffective ventilation technique (mixture of in-room or recycled air, poor mechanical ventilation servicing) have been associated with deteriorated health conditions for high-density building occupants [11].

The probability of infection from airborne disease or mitigation spread through ventilation [34] depends on quanta (viruses released), people exposure time, occupancy, activities, room ventilation flow rate, and room volume.

With the unprecedented COVID-19 outbreak sweeping the globe, many academics and industry professionals are raising the topic of whether present ventilation strategies are outmoded and inadequate for such a contagious disease in today’s environment [11, 34–36]. This prompted scientists to advocate for a paradigm shift in ventilation. There is an imperative to choose appropriate ventilation strategies and explore the way to structure ventilation systems for distributing healthy air as opposed to facilitating the accumulation of aerosol dispersal in confined environments [10, 37].

This chapter discusses an introduction to ventilation, distinct ventilation strategies such as natural ventilation, mechanical ventilation, mixed mode, or hybrid ventilation, as well as their perks and permits in preventing the spread of viruses. Also covered are the detailed case studies for analyzing mixed-mode ventilation that were published in pertinent articles after the COVID-19 pandemic.

2. Ventilation

Ventilation is the act of blending or substituting stale. Contaminated air found inside with the fresh air is drawn in from the outside of the building in order to lower the overall amount of contaminants found within it [38]. Ventilation is a crucial component for achieving high air quality and thermal comfort within a building. The installation of a ventilation system is recommended for a variety of reasons. The prime objective is to maintain the air's quality by providing an adequate amount of fresh air for metabolic processes and diluting of contaminants. A ventilation system purifies an interior space by removing impurities, undesirable heat, and excess moisture via the process of air extraction. Besides, the purposes of ventilation are (1) protecting the health and comfort of human beings, (2) in the event that it is required, providing adequate air or oxygen for activities such as combustion, (3) eliminating the byproducts of respiration as well as body odor, (4) eliminating potentially hazardous chemicals, and (5) improving air circulation [39–41].

Three fundamental aspects of building ventilation need to be taken into consideration for this design:

- *Rate of ventilation:* the quantity of outside air that is brought into the interior space, and the quality of the air that is brought in from the outdoors;
- *Airflow direction:* refers to the general direction of airflow in a building that ought to flow from clean zones to filthy zones; and
- *Air distribution or airflow pattern:* outside air ought to be supplied to each section of the space in an effective way, and airborne pollutants should be removed as they circulate around the space.

2.1 Ventilation methods

A building may be ventilated using one of the three methods: natural, mechanical, or hybrid (mixed-mode) [10, 39, 41–44]. In this chapter, we provide a briefing regarding these approaches and their potential in reducing the dissemination of the virus.

2.1.1 Natural ventilation (NV)

The act of delivering fresh air circulation inside of an enclosure by the utilization of air pressure differentials brought on largely by the effects of wind and buoyancy effect which is again caused by temperature differences in and around the enclosure is referred to as natural ventilation. This method may be defined as a natural way to provide ventilation [10, 45]. The wind that flows along the windward side eliminates the occupants' body heat through convective and evaporative heat transfer, resulting in chill body. Buoyancy effect is the effect whereby heated and less dense air rises, while cold and high-density air descends [10, 46–48]. NV is a standard green alternative to mechanical ventilation (MV) systems that contributes to energy saving as it operates without mechanical systems and energy usage by using natural forces [10, 49–51]. Two categories of NV are controlled, organized NV and nonorganied NV or infiltration [10, 52]. There are three different forms of controlled ventilation: single-sided ventilation, cross-ventilation, and passive stack ventilation [10, 49, 53]. One of the authors [46] in their earlier work has detailed about natural ventilation strategy called wind catcher. Wind catchers are an integral element of the construction process in Middle Eastern countries for many years so that buildings may take advantage of natural ventilation and cooling. The wind catcher allows for naturally occurring ventilation because of both the wind effect and the stack buoyancy effect [46]. Although the WHO strongly recommended NV as a means of reducing the risk of infection [39], it is not commonly used because of some disadvantages, including erratic airflow, a fluctuating ventilation rate, reduced thermal comfort, and it is not adequate for high occupancy levels [54–60]. As a result of these, the researchers have a hypothesis that mixed mode ventilation (MMV) or a hybrid method made up of natural and mechanical ventilation devices may be considered a more consistent strategy to offer steady airflow, diffuse contaminants, and freshen indoor air while also providing thermal comfort, particularly when paired with controls [10, 61, 62].

2.1.2 Mechanical ventilation (MV)

Mechanical ventilation systems utilize ducts and fans to circulate air. Fans can either be set straight in windows or the walls, or they can be positioned in air ducts, which allows them to either introduce fresh air into an environment or remove stale air from it [39]. Mechanical ventilation methods include distinct benefits and drawbacks. Below are some advantages [39] that may be considered more broadly:

- The integration of this system with the air conditioning system may be achieved with ease.
- The regulation of indoor humidity and temperature is readily achievable.
- The addition of a filtration system to the mechanical ventilation system is a viable option. The removal of dangerous bacteria, particles, gases, smells, and vapors may be achieved.
- Irrespective of the prevailing ambient temperature and wind conditions, continuous access to the necessary flow rate may be achieved.
- The only need for its functionality is the provision of electrical power.
- The control of airflow direction is possible.

Nevertheless, mechanical ventilation systems are not without their challenges. Mechanical ventilation systems often encounter operational challenges, leading to disruptions in their intended functionality. These interruptions may arise from several factors, such as equipment malfunction, utility service discontinuity, suboptimal design, inadequate maintenance, or improper management. In the event that the system caters to a vital facility and necessitates uninterrupted operation, it may be essential to implement backup measures for every piece of equipment. However, it is important to acknowledge that such an endeavor might incur significant costs and may not be environmentally sustainable in the long term. The expenses associated with the installation and maintenance of a mechanical ventilation system might be quite expensive. If a mechanical system is unable to be adequately built or maintained as a result of insufficient financial resources, its performance would be adversely affected [39]. Morawska et al. [63] observed the establishment of very large emergency hospital wards, such as those located inside exhibition centers, capable of accommodating hundreds or even thousands of patients. While the mechanical ventilation in these facilities is deemed sufficient for regular exhibition or conference activities, it remains uncertain whether there will be adequate ventilation for patient management and infection control when these facilities are utilized for such purposes, particularly during the COVID-19 pandemic. Studies have shown a correlation between the cleanliness of air filters and HVAC systems and the extent of symptoms tied to sick building syndromes [11]. Moreover, it has been established that inadequate ventilation rates and improper ventilation strategies, such as the mixing of in-room or recirculated air and poor maintenance of mechanical ventilation systems, have been associated with adverse health effects among those inhabiting high occupancy buildings [11]. In their study, Sha et al. [64] conducted an analysis of the rates of mechanical ventilation with the aim of mitigating the risk of COVID-19 transmission. The researchers attempted to establish a correlation between the probability of infection and the requirement for ventilation (the ventilation rate necessary is influenced by five key factors: infection probability, quantum generation rate, social distance with or without mask use, and exposure duration). When the probability of infection decreases to a certain range, there may be a significant rise in the necessary ventilation rate. As an example, while maintaining a social distance of 1.8 meters and being exposed for a duration of 8 hours, the necessary ventilation rate exhibits an increase from 2.6 air changes per hour (ACH) to 5.2 ACH as the infection chance decreases from 2–1%. Nevertheless, when the probability of infection decreases from 1 to 0.1%, there is a corresponding requirement to raise the ventilation rate from 5.2 air changes per hour (ACH) to 52.4 ACH [64]. Based on Chow's outcomes [65], it can be inferred that in order to effectively manage the spread of COVID-19 throughout various buildings, it is necessary to maintain a minimum air exchange rate of six air changes per hour (ACH). The adoption of this particular value has shown its efficacy; nevertheless, it has also presented several problems. The operational expenses associated with mechanical ventilation systems are significantly elevated. This might perhaps explain why some locations have periods of zero air changes per hour (ACH). The measurement of flow parameters in mechanical air handling systems is a time-consuming process. Determining the ventilation rate of six ACH expeditiously is a challenge. Verifying if the ventilation system is intermittently deactivated for the sake of fuel conservation presents an additional level of complexity. The transmission of viruses may occur effortlessly [65]. A research indicated the use of mechanical ventilation with recirculation as a means to minimize ventilation costs. In this particular case, the ventilation system had a role in enabling the transmission of SARS-CoV-19 to all the individuals residing in a nursing home [66, 67].

The use of mechanical ventilation systems that do not include recirculation mechanisms has been shown to mitigate the potential risk of infection [67–70]. Mechanical ventilation systems have high efficacy in managing indoor air quality (IAQ); nonetheless, it is essential to acknowledge the concomitant rise in energy consumption associated with their operation. One potentially effective strategy that might make a valuable contribution to this subject matter is the implementation of a hybrid or mixed mode ventilation technique [71, 72], which will be discussed in the next session.

2.1.3 Mixed mode or hybrid ventilation

Mixed-mode or hybrid ventilation refers to the integration of natural ventilation systems with mechanical cooling systems. Hybrid systems endeavor to maximize the utilization of free cooling via the amalgamation of outside and inside air, while using mechanical cooling to maintain thermal comfort when external circumstances are unsuitable for ventilative cooling [10, 61, 62, 71–73]. There is an increasing inclination toward enhancing research endeavors that investigate the mixed-mode ventilation concept as a means to enhance energy efficiency and improve occupant comfort in buildings [74]. Furthermore, it is well acknowledged that ventilation plays a crucial role in mitigating the transmission of COVID-19, and mixed-mode ventilation is deemed suitable for achieving this objective [75]. The following section provides a comprehensive assessment of published research studies that examine the mixed mode techniques for different buildings and the effects of design on the effective refreshment of indoor spaces, the removal of particles, and the management of indoor airborne transmission across different environments and improves the indoor air quality which aid in the mitigation of airborne viral transmission as detailed in Section 1.2.

3. Case studies

3.1 Case 1: educational classroom buildings

Ren et al. [76] provided recommendations to prevent an infectious disease spread in only naturally ventilated university classrooms. The authors suggest combining natural ventilation with window-integrated fans for hybrid ventilation and optimizing window size. Virtual simulations were done using ANSYS Fluent. The Air Diffusion Performance Index (ADPI) and Wells-Riley infection risk calculation were used to evaluate performance. It was found ADPI may be enhanced by 17% and infection risk reduced by 27%. Infection illness prevention and ventilation efficacy are greatly improved by hybrid ventilation. **Figure 1a** depicts the study classroom sketch. Rodríguez-Vidal et al. [77] compared natural, mechanical, and hybrid ventilation systems in a university classroom, analyzing their benefits and downsides in the Basque Country,

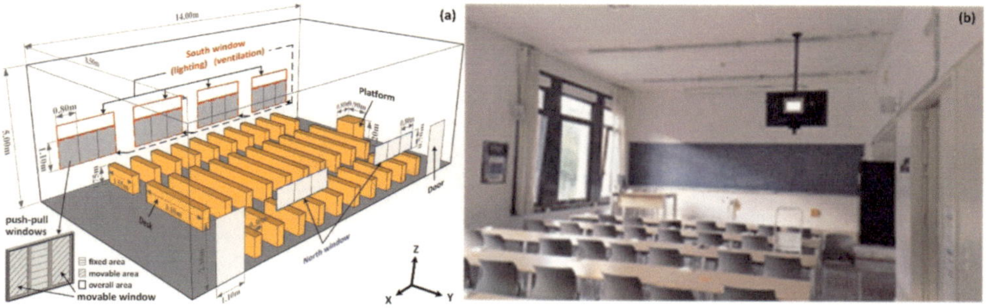

Figure 1.
Classroom model used by Ren et al. [76] (a) and Rodríguez-Vidal et al. [77] (b).

Spain. CO_2 concentrations are compared among scenarios. This research examines IAQ, including CO_2 concentration, thermal comfort, and energy analysis. For this research, 50% room occupancy was considered. The study was done using the Design Builder tool. The hybrid ventilation system efficiently controlled IAQ, creating a healthy and pleasant atmosphere for inhabitants. The hybrid system showed enhanced energy efficiency over mechanical ventilation while preserving comfortable indoor air. It improves indoor air quality and is recommended to decrease viral transmission. **Figure 1b** depicts the simulated classroom.

Quijada et al. [78] investigated how to use hybrid ventilation for university classrooms in Panama City to improve students' thermal comfort and indoor air quality. DesignBuilder 6.0 was utilized. A hybrid approach that increases fresh air intake from 5 L/s/person to 10 L/s/person via mechanical ventilation and occupancy reduction is advised. Cheong et al. [79] tested numerous ventilation techniques to reduce instructor-transmitted airborne illnesses in elementary classrooms. Traditional mechanical and hybrid ventilation solutions have been tested. This investigation uses Star CCM + CFD software. The hybrid ventilation approach that was presented showed the potential to effectively eliminating airborne infections.

3.2 Case 2: hospitals

Anuraghava et al. [80] investigated the spread of airborne viruses within a negative pressure room using a mixed-mode ventilation system. Negative-pressure isolation rooms are intended to keep hazardous particles from spreading into the surrounding environment. This is accomplished by creating a negative pressure within the space, so that when the door is opened, the ambient air is sucked in rather than the air already presents inside the room leaving. The computational fluid dynamics (CFD) tool ANSYS FLUENT was used for numerical study. The researchers used discrete-phase modeling to replicate the movement and dispersion of virus droplets while changing the ventilation flow rate parameters. The model consists of a room with two beds. There are also two human bodies on the beds, as well as two rectangular inlets and two circular outlets. **Figure 2a** displays the model under evaluation from an aerial viewpoint. The K-epsilon turbulence model was applied. The results showed that the mixed-mode ventilation system is more effective in managing virus droplet dispersion within the enclosed environment. It should be noted, however, that this research did not include a comparison with other alternative ventilation strategies. Yu et al. [81] studied the efficiency of ventilation design options in normal hospital wards for virus elimination. CFD tool ANSYS FLUENT was used to simulate

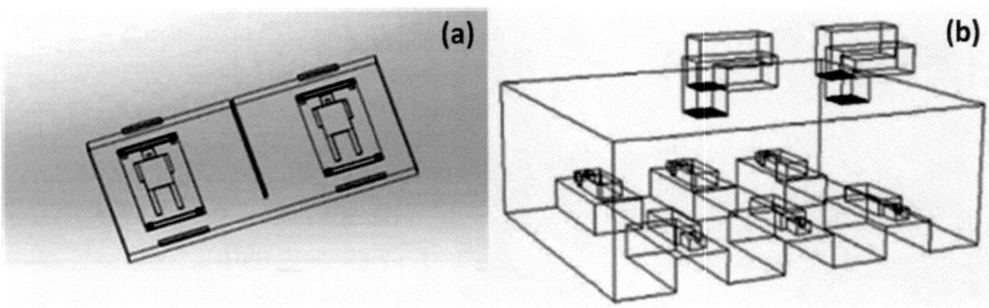

Figure 2.
Model overview of Anuraghava et al. [80] (a) and Yu et al. [81] (b).

airflow field and virus dispersion in a typical six-bed general ward in a Hong Kong hospital. The respiratory viruses included MERS-CoV, SARS-CoV, and H1N1 influenza virus. A 9 h-1 air change rate efficiently reduces respiratory virus particle deposition and floating time while optimizing energy efficiency. Despite not focusing on mixed-mode ventilation, this research on general ventilation techniques for improving IAQ may help hospital management reduce infection risk via improved ventilation design strategies. **Figure 2b** displays the model used for CFD analyses. Biological particles that may lead to surgical site infections were investigated by Liu et al. [82] in the context of the operating room (OR) air environment. The best ventilation system for maintaining clean air in the OR was discovered to be hybrid ventilation systems with the temperature-controlled airflow (TAF) system included.

3.3 Case 3: office buildings

Srivastava et al. [83] examined the effects of adding an air disinfection system to an office building's mixed-mode ventilation system. Increased air changes per hour (ACH) in confined spaces may reduce infection risk. However, this strategy significantly raises HVAC system running costs. The same goal may be achieved using region-specific air disinfection technologies. The research employed RM3 UV-C air disinfection. A CFD numerical simulation was performed on a real-life office building with offices and workstations. The model is in **Figure 3a**. This research used ANSYS FLUENT software and the RNG $k - \varepsilon$ turbulence model for CFD simulation. The Eulerian approach was chosen over the Lagrangian method for viral concentration modeling for more accurate prediction. Four cases were studied. The numerical results suggest that a hybrid system with 100% outdoor air intake and UV-C technology in HVAC ducts is ideal. This arrangement disinfects HVAC air using RM3 UV-C devices. In this research, infection risk dropped from 27% (10% outside air) to 3.1% utilizing 100% outdoor air. If infection must be reduced below 2%, it is recommended to utilize UV-C devices. Cai et al. [84] examined mixing ventilation options for upgrading an office meeting room to reduce airborne infectious virus and particle concentrations. Infection risk has been demonstrated using the Well-Riley model. The base scenario was analyzed by measuring CO_2 concentrations and particulates in the meeting room with one, two, or no individuals. Phoenics CFD tool was used to analyze improvement models. With a low-pressure filter with 99.9% efficiency and mixing ventilation, CFD models suggest an infection risk < 1%. **Figure 3b** shows the virtual model. Duan et al. [85] modeled mixed-mode ventilation to analyze the energy savings of upgrading the ventilation strategy of an office in Beijing. DesignBuilder 7.0 and CFD were used for

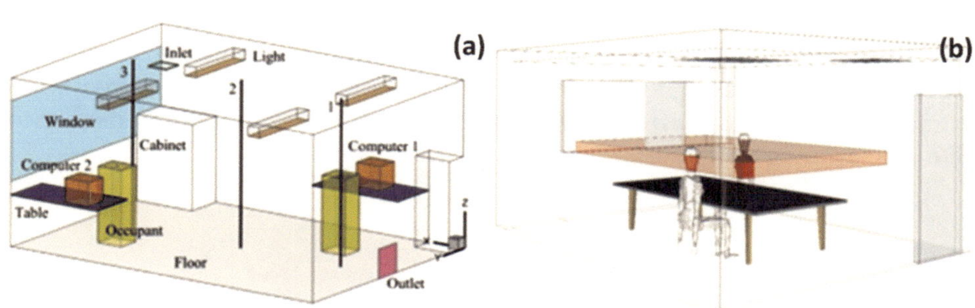

Figure 3.
Illustration of models used by Srivastava et al. [83] (a) and Cai et al. [84] (b).

the research. Hamdy and Mauro [86] presented three hybrid ventilation control technologies for an open-plan office building in Glasgow, Scotland. The three solutions were compared to a mechanical ventilation system using IDA ICE software. For indoor comfort, performance assessment includes interior temperature and predicted the percentage of dissatisfied (PPD). Assessment of indoor air quality (IAQ) includes CO_2 monitoring. Though these last two works do not involve the assessment of infectious risk, they provide good insights into optimizing the mixed mode ventilation to improve the IAQ.

3.4 Case 4: residential buildings

The study done by Al-Hilfi et al. [87] focused on the improvement of indoor air quality (IAQ) via the management of an environmental-controlled fan. The research was carried out in Malaysia. In tropical nations such as Malaysia, it is insufficient for air quality improvement to rely only on natural ventilation. However, the implementation of home isolation measures during the lockdown period has resulted in a surge in the use of air conditioning systems. This has therefore heightened individuals' susceptibility to respiratory ailments and compromised their immune system, rendering them more susceptible to the COVID-19 pandemic. The concept of designing automated fan systems that monitor indoor air quality (IAQ) in order to enhance air velocity and provide sufficient ventilation has been motivated by several environmental elements, including humidity levels, airflow velocity, concentrations of CO_2 and CO, and temperature levels. The study included the collection of data from the master bedroom of a residential unit in Malaysia over a period of about 2 weeks. Various sensors were used to measure the characteristics indicated above. The analysis of the thermal comfort state was conducted using the predicted mean vote (PMV) approach. The benchmark outlines the ten primary circumstances including all potential indoor air conditions and their associated fan speeds required in accordance with the ASHRAE 55-2020 and EN-16798 standards. While ensuring the desired level of IAQ by implementing an environmental fan system, it is also found that a significant decrease of 31.4% in energy consumption of air conditioning units is accomplished. This research primarily focuses on the optimization of hybrid ventilation systems via the use of sensors and controllers, with a specific emphasis on ensuring the maintenance of sufficient indoor air quality (IAQ) and also energy savings. Tognon et al. [88] used a co-simulation methodology to examine two distinct case studies, and one of the studies is residential building. The objective was to assess various control techniques for hybrid ventilation systems, focusing on their impact on the mitigation of risks and also energy savings. The simulations were conducted with the TRNSYS and CONTAM software, with the subsequent use of the Wells-Riley model to predict the risk of airborne infection. The study determined that when properly managed using an appropriate management method, the hybrid ventilation system shows potential in effectively sustaining indoor settings that promote health, while also lowering energy usage.

4. Conclusions

The paramount significance of appropriate ventilation strategies during periods of epidemic crises is indisputable. In this book chapter, we have undertaken an exhaustive examination of multiple dimensions pertaining to the operation of mixed-mode ventilation strategies in such challenging contexts. Our central goal is to explore the full spectrum of ventilation techniques that can be deployed to mitigate the transmission of airborne diseases.

A core theme of our discussion centers on the critical role that indoor air quality (IAQ) assumes in safeguarding health and well-being, more specifically within the framework of airborne infections during epidemics. The subject of IAQ is complex and multifaceted, involving a myriad of components, each contributing to the overall air quality experienced by individuals inside buildings.

We shed light on the importance of diverse ventilation strategies as effective tools to manage IAQ. Our examination extends to an in-depth survey of ventilation standards, both those particular to Singapore and those recognized internationally. Through this survey, we underscore the imperative of adhering to these benchmarks in striving for and sustaining optimal IAQ.

In this chapter, various ventilation strategies, such as natural, mechanical, and hybrid ventilation strategies are discussed. However, the primary emphasis of this chapter lies in the domain of mixed-mode or hybrid ventilation. As an innovative approach that amalgamates the merits of natural and mechanical ventilation techniques, mixed-mode ventilation emerges as a plausible and potentially energy-efficient solution to manage IAQ efficaciously. An extensive dissection of this technique is provided, delineating its principles, potential advantages, and the contexts in which it can be aptly applied.

To supplement our analysis, we delve into a range of case studies that portray the practical applications of the aforementioned ventilation strategies. These case studies elucidate the implementation and optimization of these techniques in real-world settings, further emphasizing the effectiveness of mixed-mode ventilation across different environments. Our case studies include various building types, such as educational buildings, hospitals, office buildings, and residential buildings. Each case study employs a unique analytical approach, ranging from computational fluid dynamics (CFD) to empirical experimentation. This diverse set of methodologies provides a holistic picture of the challenges and opportunities involved in improving IAQ across different environments.

In conclusion, robust ventilation strategies, and in particular mixed-mode ventilation, are instrumental in sustaining IAQ, mitigating the risk of airborne infections, and augmenting occupant comfort, particularly during epidemic crises. However, this chapter is a stepping stone in this ever-evolving field. Further research is warranted to foster the development of robust, adaptive strategies for the effective implementation of mixed-mode ventilation. Future investigations should aim to enhance energy efficiency, comfort, and safety across a diverse portfolio of building types and environmental conditions. In another chapter *Machine Learning Techniques in Indoor Environmental Quality Assessment,* we have discussed recent advancements in ML-based techniques for Indoor Environmental Quality Assessment.

Author details

Ijaz Fazil Syed Ahmed Kabir[1†], Mohan Kumar Gajendran[2†] and Eddie Yin-Kwee Ng[1*]

1 School of Mechanical and Aerospace Engineering, Nanyang Technological University, Singapore

2 Division of Energy, Matter and Systems, University of Missouri-Kansas city, Kansas City, MO, USA

*Address all correspondence to: mykng@ntu.edu.sg

† These authors contributed equally.

References

[1] Mendes A, Teixeira JP. Sick building syndrome. In: Wexler P, editor. Encyclopedia of Toxicology. 3rd ed. Oxford: Academic Press; 2014. pp. 256-260. DOI: 10.1016/B978-0-12-386454-3.00432-2. ISBN: 978-0-12-386455-0

[2] Güneş G, Yalçin N, Çolaklar H. Investigation of indoor air quality in university libraries in terms of gaseous and particulate pollutants in Bartin, Turkey. Environmental Monitoring and Assessment. 2022;**194**(3):200. DOI: 10.1007/s10661-022-09818-8. ISSN: 1573-2959

[3] Abdul-Wahab SA, En SCF, Elkamel A, Ahmadi L, Yetilmezsoy K. A review of standards and guidelines set by international bodies for the parameters of indoor air quality. Atmospheric Pollution Research. 2015;**6**(5):751-767. DOI: 10.5094/APR.2015.084. ISSN: 1309-1042

[4] Kubba S. Handbook of Green Building Design and Construction: LEED, BREEAM, and Green Globes. Oxford, UK: Butterworth-Heinemann/Elsevier; 2012

[5] Soreanu G. 12 - Biotechnologies for improving indoor air quality. In: Pacheco-Torgal F, Rasmussen E, Granqvist C-G, Ivanov V, Kaklauskas A, Makonin S, editors. Start-Up Creation. Oxford, UK: Woodhead Publishing/Elsevier; 2016. pp. 301-328. DOI: 10.1016/B978-0-08-100546-0.00012-1. ISBN: 978-0-08-100546-0

[6] ANSI ASHRAE. Ashrae standard 62.1-2022. In: Ventilation and Acceptable Indoor Air Quality. Atlanta, GA: American Society of Heating, Refrigerating, and Air-Conditioning Engineers, Inc.; 2021

[7] Roh T, Moreno-Rangel A, Baek J, Obeng A, Hasan NT, Carrillo G. Indoor air quality and health outcomes in employees working from home during the COVID-19 pandemic: A pilot study. Atmosphere. 2021;**12**(12). DOI: 10.3390/atmos12121665. ISSN: 2073-4433

[8] Nag PK. Sick Building Syndrome and Other Building-Related Illnesses. Singapore: Springer Singapore; 2019. pp. 53-103. DOI: 10.1007/978-981-13-2577-9_3. ISBN: 978-981-13-2577-9

[9] Arjmandi H, Amini R, Kashfi M, Abikenari MA, Davani A. Minimizing the COVID-19 spread in hospitals through optimization of ventilation systems. Physics of Fluids. 2022;**34**(3): 037103. DOI: 10.1063/5.0081291. ISSN: 1070-6631

[10] Izadyar N, Miller W. Ventilation strategies and design impacts on indoor airborne transmission: A review. Building and Environment. 2022;**218**: 109158. DOI: 10.1016/j.buildenv.2022.109158. ISSN: 0360-1323

[11] Lipinski T, Ahmad D, Serey N, Jouhara H. Review of ventilation strategies to reduce the risk of disease transmission in high occupancy buildings. International Journal of Thermofluids. 2020;**7-8**:100045. DOI: 10.1016/j.ijft.2020.100045. ISSN: 2666-2027

[12] World Health Organization. Statement on the second meeting of the international health regulations (2005) emergency committee regarding the outbreak of novel coronavirus (2019-ncov). 2020a [Accessed: July 28, 2023]

[13] World Health Organization. Novel coronavirus (2019-ncov): Situation report, 11. Technical documents [Accessed: Jan 31, 2020]

[14] World Health Organization. Coronavirus disease (COVID-19) situation report – 152. Technical report. World Health Organization. 2020b. Available from: https://www.who.int/docs/default-source/coronaviruse/situation-reports/20200620-covid-19-sitrep-152.pdf [Accessed: July 28, 2023]

[15] Wang D, Zhou M, Nie X, Qiu W, Yang M, Wang X, et al. Epidemiological characteristics and transmission model of corona virus disease 2019 in China. Journal of Infection. 2020;**80**(5):e25-e27

[16] Wu Z, McGoogan JM. Characteristics of and important lessons from the coronavirus disease 2019 (COVID-19) outbreak in China: Summary of a report of 72314 cases from the Chinese Center for Disease Control and Prevention. JAMA. 2020;**323**(13): 1239-1242. DOI: 10.1001/jama.2020.2648. ISSN: 0098-7484

[17] Guo Z-D, Wang Z-Y, Zhang S-F, Li X, Li L, Li C, et al. Aerosol and surface distribution of severe acute respiratory syndrome coronavirus 2 in hospital wards, Wuhan, China, 2020. Emerging Infectious Diseases. 2020;**26**(7): 1583-1591

[18] Mittal R, Ni R, Seo J-H. The flow physics of COVID-19. Journal of Fluid Mechanics. 2020;**894**:F2

[19] Gameiro M, da Silva. An analysis of the transmission modes of COVID-19 in light of the concepts of indoor air quality. REHVA Journal. 2020;**3**:46-54

[20] Almaimani A, Alaidroos A, Krarti M, Qurnfulah E, Tiwari A. Evaluation of optimal mechanical ventilation strategies for schools for reducing risks of airborne viral infection. Buildings. 2023;**13**(4). DOI: 10.3390/buildings13040871. ISSN: 2075-5309

[21] Berry G, Parsons A, Morgan M, Rickert J, Cho H. A review of methods to reduce the probability of the airborne spread of COVID-19 in ventilation systems and enclosed spaces. Environmental Research. 2022;**203**: 111765. DOI: 10.1016/j.envres.2021.111765. ISSN: 0013-9351

[22] Schumacher S, Sanchez AB, Caspari A, Staack K, Asbach C. Testing filter-based air cleaners with surrogate particles for viruses and exhaled droplets. Atmosphere. 2022;**13**(10). DOI: 10.3390/atmos13101538. ISSN: 2073-4433

[23] Tang JW, Li Y, Eames I, Chan PKS, Ridgway GL. Factors involved in the aerosol transmission of infection and control of ventilation in healthcare premises. Journal of Hospital Infection. 2006;**64**(2):100-114. DOI: 10.1016/j.jhin.2006.05.022. ISSN: 0195-6701

[24] Bhagat RK, Davies Wykes MS, Dalziel SB, Linden PF. Effects of ventilation on the indoor spread of COVID-19. Journal of Fluid Mechanics. 2020;**903**:F1. DOI: 10.1017/jfm.2020.720

[25] Li Y, Leung GM, Tang JW, Yang X, Chao CYH, Lin JZ, et al. Role of ventilation in airborne transmission of infectious agents in the built environment – A multidisciplinary systematic review. Indoor Air. 2007; **17**(1):2-18. DOI: 10.1111/j.1600-0668.2006.00445.x. ISSN: 0905-6947

[26] Aernout Somsen G, van Rijn C, Kooij S, Bem RA, Bonn D. Small droplet aerosols in poorly ventilated spaces and SARS-COV-2 transmission.

The Lancet Respiratory Medicine. 2020; **8**(7):658-659

[27] Nishiura H, Oshitani H, Kobayashi T, Saito T, Sunagawa T, Matsui T, et al. Closed environments facilitate secondary transmission of coronavirus disease 2019 (COVID-19). medRxiv. 2020. DOI: 10.1101/2020.02.28.20029272

[28] Carrer P, Wargocki P, Fanetti A, Bischof W, Fernandes EDO, Hartmann T, et al. What does the scientific literature tell us about the ventilation–health relationship in public and residential buildings? Building and Environment. 2015;**94**:273-286. DOI: 10.1016/j.buildenv.2015.08.011. ISSN: 0360-1323

[29] World Health Organization. Regional Office for Europe. Development of who guidelines for indoor air quality: Report on a working group meeting, Bonn, Germany, 23–24 October 2006. Technical documents. 2006

[30] World Health Organization et al. Clinical management of severe acute respiratory infection (SARI) when COVID-19 disease is suspected: Interim guidance. Paediatrics and Family Medicine. 2020;**16**(1):9

[31] World Health Organization. Novel coronavirus 2019, advice for public. Available from: https://www.who.int/emergencies/diseases/novel-coronavirus-2019/advice-for-public

[32] World Health Organization. Modes of transmission of virus causing COVID-19: Implications for IPC precaution recommendations: Scientific brief. Technical documents. 2020c [Accessed: March 27, 2020]

[33] World Health Organization, et al. Infection prevention and control during health care when coronavirus disease (COVID-19) is suspected or confirmed: Interim guidance. Technical report. World Health Organization. 2021 [Accessed: July 12, 2021]

[34] Rivas E, Santiago JL, Martín F, Martilli A. Impact of natural ventilation on exposure to SARS-COV 2 in indoor/semi-indoor terraces using CO_2 concentrations as a proxy. Journal of Building Engineering. 2022;**46**:103725. DOI: 10.1016/j.jobe.2021.103725. ISSN: 2352-7102

[35] Benito PA, Ramírez CP, Viccione G, Lepore E. Ventilation for residential buildings: Critical assessment of standard requirements in the COVID-19 pandemic context. Frontiers in Built Environment. 2021;7:656718

[36] Rencken GK, Rutherford EK, Ghanta N, Kongoletos J, Glicksman L. Patterns of SARS-COV-2 aerosol spread in typical classrooms. Building and Environment. 2021;**204**:108167. DOI: 10.1016/j.buildenv.2021.108167. ISSN: 0360-1323

[37] Melikov AK. COVID-19: Reduction of airborne transmission needs paradigm shift in ventilation. Building and Environment. 2020;**186**:107336. DOI: 10.1016/j.buildenv.2020.107336. ISSN: 0360-1323

[38] Awb HB. Ventilation for good indoor air quality and energy efficiency. Energy Procedia. 2017;112:277-286. DOI: 10.1016/j.egypro.2017.03.1098. Available from: https://www.sciencedirect.com/science/article/pii/S1876610217312237. ISSN: 1876-6102

[39] Atkinson J, Chartier Y, Pessoa-Silva CL, Jensen P, Li Y, Seto W-H. Natural Ventilation for Infection Control in Health-Care Settings. Geneva, Switzerland: WHO; 2009a. Available

from: https://www.who.int/water_sanitation_health/publications/natural_ventilation.pdf

[40] Awbi HB. Ventilation of Buildings. 2nd ed. London: Routledge; 2003. DOI: 10.4324/9780203634479. ISBN: 9780203634479 [Accessed: June 4, 2003]

[41] Yang T, Clements-Croome DJ. Natural Ventilation in Built Environment. US, New York, NY: Springer; 2020. pp. 431-464. DOI: 10.1007/978-1-0716-0684-1_488. ISBN: 978-1-0716-0684-1

[42] Levin H. Natural Ventilation: Theory. Santa Cruz, California, US: Building Ecology Research Group; 2009

[43] Etheridge DW, Sandberg M. Building Ventilation: Theory and Measurement. Vol. 50. Chichester, UK: John Wiley & Sons; 1996

[44] Bolashikov ZD, Melikov AK. Methods for indoor air disinfection and purification from airborne pathogens for application in HVAC systems. In: Proceedings of the 6th International Conference on Indoor Air Quality, Ventilation and Energy Conservation in Buildings, Citeseer, Sendai, Japan. Vol. 2. 2007. pp. 565-573

[45] Whittleton D. Industrial buildings. In: Plant Engineer's Reference Book. Oxford, UK: Elsevier; 1991. pp. 2-1

[46] Benkari N, Fazil I, Husain A. Design and performance comparison of two patterns of wind-catcher for a semi-enclosed courtyard. International Journal of Mechanical Engineering and Robotics Research. 2017;**6**(5):396-400

[47] Larsen TS. Natural ventilation driven by wind and temperature difference [PhD thesis] Denmark. Vedlagt dokument tilpasset A4 format - 134 sider (140 siders print). 2006

[48] Larsen TS, Heiselberg P. Single-sided natural ventilation driven by wind pressure and temperature difference. Energy and Buildings. 2008;**40**(6): 1031-1040. DOI: 10.1016/j.enbuild.2006.07.012. ISSN: 0378-7788

[49] Izadyar N, Miller W, Rismanchi B, Garcia-Hansen V. Impacts of façade openings' geometry on natural ventilation and occupants' perception: A review. Building and Environment. 2020;**170**:106613. DOI: 10.1016/j.buildenv.2019.106613. ISSN: 0360-1323

[50] Hughes BR, Chaudhry HN, Ghani SA. A review of sustainable cooling technologies in buildings. Renewable and Sustainable Energy Reviews. 2011;**15**(6):3112-3120. DOI: 10.1016/j.rser.2011.03.032. ISSN: 1364-0321

[51] Alan Short C, Al-Maiyah S. Design strategy for low-energy ventilation and cooling of hospitals. Building Research & Information. 2009;**37**(3):264-292. DOI: 10.1080/09613210902885156

[52] Melikov A, K, Markov DG. Novel approach for evaluation of air change rate in naturally ventilated occupied spaces based on metabolic CO_2 time variation. In: Proceedings of Indoor Air 2014. 13th International Conference on Indoor Air Quality and Climate; Conference Date: 07-07-2014 through 12-07-2014. Hong Kong: International Society of Indoorair Quality and Climate (ISIAQ); 2014. Available from: http://www.indoorair2014.org/

[53] Ghiaus C, Roulet C-A. Strategies for Natural Ventilation. 1st ed. Abingdon, UK: Routledge/Taylor & Francis; 2005.

DOI: 10.4324/9781849772068-7. ISBN: 9781849772068

[54] Sahu AK, Verma TN, Sinha SL. Numerical simulation of air flow in multiple beds intensive care unit of hospital. International Journal of Automotive and Mechanical Engineering. 2019;**16**(2):6796-6807. DOI: 10.15282/ijame.16.2.2019.24.0511

[55] Ahmed T, Kumar P, Mottet L. Natural ventilation in warm climates: The challenges of thermal comfort, heatwave resilience and indoor air quality. Renewable and Sustainable Energy Reviews. 2021;**138**:110669. DOI: 10.1016/j.rser.2020.110669. ISSN: 1364-0321

[56] Yadav AK, Maurya C. An environmental quality assessment: Study on natural ventilation in buildings. International Research Journal of Engineering and Technology (IRJET). 2019;**6**:1801-1805. Available from: https://www.irjet.net/archives/V6/i5/IRJET-V6I5228.pdf. ISSN: 2395-0056. ISO 9001:2008 Certified Journal

[57] Stockwell RE, Ballard EL, O'Rourke P, Knibbs LD, Morawska L, Bell SC. Indoor hospital air and the impact of ventilation on bioaerosols: A systematic review. Journal of Hospital Infection. 2019;**103**(2):175-184. DOI: 10.1016/j.jhin.2019.06.016. ISSN: 0195-6701

[58] Cao G-Q, Zhang Y-Z. The impact of ventilation and air filtration in controlling indoor airborne microbes. Tumu Jianzhu yu Huanjing Gongcheng/Journal of Civil, Architectural and Environmental Engineering. 2009;**31**: 130-140

[59] Huang K, Song J, Feng G, Chang Q, Jiang B, Wang J, et al. Indoor air quality analysis of residential buildings in Northeast China based on field measurements and longtime monitoring. Building and Environment. 2018;**144**: 171-183. DOI: 10.1016/j.buildenv.2018.08.022. ISSN: 0360-1323

[60] Fonseca A, Abreu I, Guerreiro MJ, Abreu C, Silva R, Barros N. Indoor air quality and sustainability management—Case study in three Portuguese healthcare units. Sustainability. 2019;**11**(1). DOI: 10.3390/su11010101. ISSN: 2071-1050.

[61] Belmans B, Aerts D, Verbeke S, Audenaert A, Descamps F. Set-up and evaluation of a virtual test bed for simulating and comparing single- and mixed-mode ventilation strategies. Building and Environment. 2019;**151**: 97-111. DOI: 10.1016/j.buildenv.2019.01.027. ISSN: 0360-1323

[62] Daaboul J, Ghali K, Ghaddar N. Mixed-mode ventilation and air conditioning as alternative for energy savings: A case study in Beirut current and future climate. Energy Efficiency. 2018;**11**:13-30. DOI: 10.1007/s12053-017-9546-z. ISSN: 1570-6478

[63] Morawska L, Tang JW, Bahnfleth W, Bluyssen PM, Boerstra A, Buonanno G, et al. How can airborne transmission of COVID-19 indoors be minimised? Environment International. 2020;**142**: 105832. DOI: 10.1016/j.envint.2020.105832. ISSN: 0160-4120

[64] Sha H, Zhang X, Qi D. Optimal control of high-rise building mechanical ventilation system for achieving low risk of COVID-19 transmission and ventilative cooling. Sustainable Cities and Society. 2021;**74**:103256. DOI: 10.1016/j.scs.2021.103256. ISSN: 2210-6707

[65] Chow WK. Ventilation requirements and recommendations for controlling SARS-COV-2 and variants outbreaks in

indoor gathering places with close contact. medRxiv. 2022. DOI: 10.1101/2022.06.15.22276447

[66] de Man P, Paltansing S, Ong DSY, Vaessen N, van Nielen G, Koeleman JGM. Outbreak of coronavirus disease 2019 (COVID-19) in a nursing home associated with aerosol transmission as a result of inadequate ventilation. Clinical Infectious Diseases. 2020;**73**(1):170-171. DOI: 10.1093/cid/ciaa1270. ISSN: 1058-4838

[67] Ratajczak K. Ventilation strategy for proper IAQ in existing nurseries buildings - lesson learned from the research during COVID-19 pandemic. Aerosol and Air Quality Research. 2022; **22**(3):210337. DOI: 10.4209/aaqr.210337

[68] Almilaji O. Air recirculation role in the spread of COVID-19 onboard the diamond princess cruise ship during a quarantine period. Aerosol and Air Quality Research. 2021;**21**(4):200495. DOI: 10.4209/aaqr.200495

[69] Li H, Shankar SN, Witanachchi CT, Lednicky JA, Loeb JC, Mahbubul Alam M, et al. Environmental surveillance and transmission risk assessments for SARS-CoV-2 in a fitness center. Aerosol and Air Quality Research. 2021;**21**(11):210106. DOI: 10.4209/aaqr.210106

[70] Schibuola L, Tambani C. High energy efficiency ventilation to limit COVID-19 contagion in school environments. Energy and Buildings. 2021;**240**:110882. DOI: 10.1016/j.enbuild.2021.110882. ISSN: 0378-7788

[71] Zaniboni L, Albatici R. Natural and mechanical ventilation concepts for indoor comfort and well-being with a sustainable design perspective: A systematic review. Buildings. 2022;**12**(11). DOI: 10.3390/buildings12111983. ISSN: 2075-5309

[72] Tognon G, Marigo M, De Carli M, Zarrella A. Mechanical, natural and hybrid ventilation systems in different building types: Energy and indoor air quality analysis. Journal of Building Engineering. 2023;**76**:107060. ISSN: 23527102

[73] Gomis LL, Fiorentini M, Daly D. Potential and practical management of hybrid ventilation in buildings. Energy and Buildings. 2021;**231**:110597. DOI: 10.1016/j.enbuild.2020.110597. ISSN: 0378-7788

[74] Kim J, de Dear R. Is mixed-mode ventilation a comfortable low-energy solution? A literature review. Building and Environment. 2021;**205**:108215. DOI: 10.1016/j.buildenv.2021.108215. ISSN: 0360-1323

[75] Arata S, Kawakubo S. Study on productivity of office workers and power consumption of air conditioners in a mixed-mode ventilation building during springtime. Building and Environment. 2022;**214**:108923. DOI: 10.1016/j.buildenv.2022.108923. ISSN: 0360-1323

[76] Ren C, Cao S-J, Haghighat F. A practical approach for preventing dispersion of infection disease in naturally ventilated room. Journal of Building Engineering. 2022;**48**:103921

[77] Rodríguez-Vidal I, Otaegi J, Armengod U. Evaluating the feasibility of hybrid ventilation in educational spaces: A simulation study in the Basque country climate. Environmental and Climate Technologies. 2023;**27**(1): 532-544

[78] Quijada MC, Solano T, Austin MC. Optimal hybrid ventilation strategy to

assure adequate indoor thermal comfort and air quality in educational spaces under a tropical climate. In: Journal of Physics: Conference Series. Vol. 2385. Bristol, UK: IOP Publishing; 2022. p. 012095

[79] Cheong CH, Hwang S-H, Park B. Ventilation methods to prevent the spread of airborne pathogens from teachers to students. International Journal of Air-Conditioning and Refrigeration. 2023;**31**(1):9

[80] Anuraghava C, Abhiram K, Reddy VNS, Rajan H. CFD modelling of airborne virus diffusion characteristics in a negative pressure room with mixed mode ventilation. International Journal for Simulation and Multidisciplinary Design Optimization. 2021;**12**:1

[81] Yu HC, Mui KW, Wong LT, Chu HS. Ventilation of general hospital wards for mitigating infection risks of three kinds of viruses including middle east respiratory syndrome coronavirus. Indoor and Built Environment. 2017;**26**(4):514-527

[82] Liu Z, Liu H, Yin H, Rong R, Cao G, Deng Q. Prevention of surgical site infection under different ventilation systems in operating room environment. Frontiers of Environmental Science & Engineering. 2021;**15**:1-14

[83] Srivastava S, Zhao X, Manay A, Chen Q. Effective ventilation and air disinfection system for reducing coronavirus disease 2019 (COVID-19) infection risk in office buildings. Sustainable Cities and Society. 2021;**75**: 103408

[84] Wenyan C, Bokel RMJ, van den Engel P. How to Reduce COVID-19 Transmission in a Small Meeting Room Using a Mixed Ceiling Ventilation System. Belgium: Air Infiltration and Ventilation Centre (AIVC); 2022

[85] Duan Z, Sun Y, Wang M, Ran H, Dong X. Evaluation of mixed-mode ventilation thermal performance and energy saving potential from retrofitting a Beijing office building. Buildings. 2022; **12**(6):793

[86] Hamdy M, Mauro GM. Optimizing hybrid ventilation control strategies toward zero-cooling energy building. Frontiers in Built Environment. 2019; **5**:97

[87] Al-Hilfi LMA, Wong J, Lim YS, Chua KH. Evaluating energy saving potential and IAQ improvement of an environmental-controlled fan system for residence during the COVID-19 era. In: 2021 9th International Conference on Smart Grid and Clean Energy Technologies (ICSGCE). Piscataway, NJ, USA: IEEE; 2021. pp. 116-122

[88] Tognon G, Marigo M, De Carli M, Zarrella A. Mechanical, natural and hybrid ventilation systems in different building types: Energy and indoor air quality analysis. Journal of Building Engineering. 2023;**76**:107060. DOI: 10.1016/j.jobe.2023.107060. Available from: https://www.sciencedirect.com/science/article/pii/S2352710223012391. ISSN: 2352-7102

Chapter 5

Machine Learning Techniques in Indoor Environmental Quality Assessment

Mohan Kumar Gajendran, Ijaz Fazil Syed Ahmed Kabir, Sudhakar Vadivelu, Eddie Yin-Kwee Ng and Ravi Chandra Thota

Abstract

This chapter provides a comprehensive exploration of the evolving role of machine learning in Indoor Environmental Quality (IEQ) assessment. As urban living spaces become increasingly enclosed, the importance of maintaining optimal IEQ for human health and well-being has surged. Traditional methods for IEQ assessment, while effective, often fail to provide real-time monitoring and control. This gap is increasingly being addressed by the integration of machine learning techniques, allowing for enhanced predictive modeling, real-time optimization, and robust anomaly detection. The chapter delves into a comparative analysis of various machine learning techniques including supervised, unsupervised, and reinforcement learning, demonstrating their unique benefits in IEQ assessment. Practical implementations of these techniques in residential, commercial, and specialized environments are further illustrated through detailed case studies. The chapter also addresses the existing challenges in implementing machine learning for IEQ assessment and provides an outlook on future trends and potential research directions. The comprehensive review offered in this chapter encourages continued innovation and research in leveraging machine learning. for more efficient and effective IEQ assessment.

Keywords: indoor environmental quality, machine learning, IEQ assessment, predictive modeling, supervised learning, unsupervised learning, reinforcement learning, real-time control, anomaly detection, residential buildings, commercial buildings, future trends

1. Introduction

This chapter delves into the integration of Indoor Environmental Quality (IEQ) assessment with Machine Learning (ML), emphasizing the critical role of optimal IEQ in urban living spaces. It explores the impact of IEQ on occupant health, comfort, and productivity [1], and examines how ML can revolutionize IEQ assessment, offering significant advantages for building management, urban planning, and public health sectors.

IEQ encompasses the conditions within buildings that affect human health and comfort, including air quality, thermal comfort, acoustical comfort, and visual comfort [2]. These elements, as illustrated in **Figure 1**, significantly influence occupant well-being and productivity [4]. The scope of IEQ spans various building types, from residential to commercial and institutional spaces like schools and hospitals [5]. Monitoring of IEQ involves diverse methods, ranging from manual controls to automated systems, and adheres to standards set by organizations like the EPA and ASHRAE [6].

Accurate and efficient assessment of IEQ is essential for maintaining healthy indoor environments. It aids in optimizing HVAC systems for energy efficiency and comfort [7] and addresses health concerns such as Sick Building Syndrome (SBS) and Building-Related Illnesses (BRI) [8]. Traditional IEQ assessment methods often fall short in real-time monitoring and managing the dynamic nature of indoor environments [9].

Machine learning, a transformative branch of artificial intelligence, is increasingly applied in IEQ assessment. ML offers advanced capabilities in data processing, predictive modeling, pattern recognition, anomaly detection, and real-time adaptive control [10]. Through various techniques, such as supervised, unsupervised, and reinforcement learning, ML enhances the accuracy and efficiency of IEQ assessments [11].

IEQ is a complex, multifaceted concept integral to occupant well-being. It spans a wide range of parameters, from physical to biochemical and psychological aspects [12]. The trend towards energy-efficient building designs has intensified the need to focus on maintaining IEQ [7], which includes factors like air quality, lighting, thermal

Figure 1.
Factors Influencing IEQ [3].

conditions, and acoustics, as well as emerging considerations such as ergonomics, esthetics, and electromagnetic field quality, as defined by entities like ASHRAE, the EPA, and WHO [13–15].

Each aspect of IEQ, critically impacting human health and comfort, is depicted in **Figure 1**. For instance, poor air quality can lead to a range of health issues, from allergies to severe conditions like cancer [1]. Inadequate thermal conditions can diminish comfort and reduce productivity [2], while neglect in acoustical and visual comfort can result in stress, cognitive impairments, and long-term mental health issues [16, 17].

To illustrate further, air quality is influenced by various pollutants, both internal and external to the building, with effects ranging from minor discomfort to serious conditions like asthma and cancer [18]. Thermal comfort, a subjective but crucial metric, hinges on multiple environmental and personal factors [2]. Acoustical comfort, dependent on sound characteristics, is vital for mental well-being [16]. Visual comfort, shaped by lighting and visual elements, affects everything from eye strain to mood and circadian rhythms [17].

The integration of ML into IEQ assessment represents a significant advancement. ML's ability to analyze complex datasets and discern patterns enables a more nuanced understanding and control of indoor environments. This chapter aims to explore the applications of ML in enhancing various facets of IEQ, from air quality analysis to thermal comfort prediction. The integration of ML in IEQ assessment not only promises improved real-time monitoring and proactive management but also opens doors to personalized environmental controls tailored to individual needs and health requirements.

As we delve into the specifics of applying ML to IEQ, the chapter will address both the technical advancements and the challenges, including ethical considerations. Our goal is to provide a comprehensive overview of how ML can be leveraged to improve indoor environmental quality, ultimately contributing to the enhanced health, comfort, and well-being of occupants in urban living spaces.

2. IEQ assessment: Traditional methods and their limitations

The assessment of Indoor Environmental Quality (IEQ) plays a pivotal role in ensuring environments that support the health, comfort, and productivity of building occupants. While traditional methods of IEQ assessment have been the cornerstone of this field, they exhibit significant limitations, impacting their overall effectiveness and reliability. In this section, we will expand upon these traditional methods, underscoring their constraints and paving the way for the introduction of machine learning techniques as a more robust solution.

2.1 Traditional assessment methods

2.1.1 Spot measurements

- *Description*: This method involves taking discrete measurements of various IEQ parameters such as temperature, humidity, carbon dioxide levels, particulate matter, and light intensity at a specific moment using specialized instruments [19].

- *Advantages*: The primary benefits include simplicity in execution and the immediate availability of results, making it a convenient option for quick assessments.
- *Limitations*: However, spot measurements may not accurately represent ongoing exposure conditions, as they overlook temporal and spatial variability. The accuracy of these measurements is also heavily dependent on the calibration and precision of the instruments used [20].

2.1.2 Long-term monitoring

- *Description*: This approach utilizes sensors or data loggers to continuously record IEQ parameters over extended periods. It is designed to capture daily and seasonal variations, providing a more comprehensive view of the indoor environment [21].
- *Advantages*: Offers a broader and more detailed perspective on IEQ, capturing long-term trends and fluctuations.
- *Limitations*: The challenge lies in the significant resources required for data management and analysis. Additionally, there is a potential for data errors and inconsistencies due to sensor calibration issues or environmental interference [22].

2.1.3 Occupant surveys

- *Description*: This method involves collecting subjective responses from occupants about their perceptions and satisfaction with the indoor environment [23].
- *Advantages*: Occupant surveys can uncover issues that might be overlooked by physical measurements, providing valuable insights into the human aspect of IEQ.
- *Limitations*: However, these surveys are prone to response biases and may not accurately reflect the actual IEQ conditions. The qualitative nature of this data also makes it challenging to quantify or systematically analyze [24].

2.1.4 Physical inspections

- *Description*: Physical inspections involve thorough examinations of the building by professionals to identify visible or detectable issues such as water leaks, mold growth, or insufficient ventilation [25].
- *Advantages*: They are particularly effective in identifying overt and tangible problems within a building.
- *Limitations*: The limitation of physical inspections lies in their inability to detect latent or intermittent problems. They also require significant expertise and can be time-consuming and labor-intensive [26].

2.2 Limitations and the need for advanced methods

The traditional methods discussed above share a common drawback: their limited ability to provide real-time, continuous data. They often fail to capture the dynamic and interactive nature of various IEQ parameters, leading to an incomplete or skewed understanding of the indoor environment. Furthermore, these methods generally lack predictive capabilities, which are essential for proactive IEQ management. They can also be resource-intensive, both in terms of time and financial investment [27].

2.3 Emerging advancements and the role of machine learning

In response to these limitations, emerging advancements in sensor technology and data analytics, particularly machine learning (ML), present promising solutions. Machine learning techniques, with their capability to process vast amounts of data, enable continuous monitoring, real-time analysis, and predictive modeling of IEQ parameters. This paradigm shift aims not only to enhance the accuracy and comprehensiveness of IEQ assessments but also to streamline the process, reducing the demand on resources. The integration of ML in IEQ assessment represents a significant step towards more intelligent, adaptive, and occupant-centered building management.

However, it is important to note that the full-scale adoption of these advanced technologies is still in progress. Further research and development are necessary to establish their effectiveness, ease of use, and integration into existing building management systems. As the field of IEQ assessment evolves, the potential of machine learning to transform this domain is immense, offering a pathway towards more sustainable, healthy, and productive indoor environments [27, 28].

3. An overview of machine learning techniques

Machine learning, as a pivotal subfield of artificial intelligence, has seen notable evolution since its inception. This discipline, coined by Arthur Samuel in 1959, initially revolved around the idea of giving "computers the ability to learn without being explicitly programmed" [29]. Samuel's work with checkers-playing programs laid the groundwork for subsequent progress in this field. The following years saw pioneering efforts from researchers like Rosenblatt who, in 1958, introduced the concept of the perceptron, an early neural network that could classify linearly separable patterns [30]. At the same time, Nilsson and Minsky's explorations in learning machines added depth to the understanding of how computational models could replicate cognitive processes [31].

The 1970s and 1980s introduced the concepts of decision trees [32], a fundamental machine learning approach which paved the way for more sophisticated algorithms like Random Forests. This era also witnessed the emergence of the kernel methods, which includes the Support Vector Machine (SVM) - a powerful classification tool that marked a significant shift in the direction of machine learning [33]. The advent of the Backpropagation algorithm in the 1980s sparked a renewed interest in neural networks [34]. However, the full potential of this method wasn't realized until recent years, due to the limitation in computational capabilities and lack of large-scale, labeled datasets.

The 1990s saw a move towards data-driven approaches, prompted by the increasing availability of digital data. The introduction of ensemble methods such as boosting and bagging brought forth a new phase in machine learning, improving model robustness and predictive power [35]. The turn of the millennium marked the era of 'big data', pushing the boundaries of machine learning further. The development of more sophisticated algorithms, like deep learning, could now be propelled by the explosion of data and advancements in computational capabilities [36]. The seminal AlexNet model, developed by Krizhevsky, Sutskever, and Hinton, for the ImageNet competition in 2012, demonstrated the incredible potential of deep learning in practice, revolutionizing the field and driving a resurgence in neural networks [37].

Today, machine learning has permeated various sectors, revolutionizing healthcare, finance, environmental science, and many more fields. From a theoretical concept to a practical tool, machine learning has become an integral part of scientific research and technological development.

3.1 Fundamentals of machine learning

In this subsection, we will delve into the core concepts that form the foundation of machine learning, encompassing its definition, the nature of the data it utilizes, the problems it addresses, and the metrics used to evaluate its performance. Machine learning, a subset of artificial intelligence, enables computers to learn autonomously by training mathematical models on data, thereby going beyond the explicit programming used in traditional computing [38]. Unlike classical programming that relies on pre-defined rules, machine learning algorithms identify patterns within data, offering a unique capacity for experiential learning.

Data is essential for machine learning performance. It can vary in type—numerical, categorical, or a blend—and often requires preprocessing steps such as data cleaning and missing data imputation to ensure model quality [38]. Machine learning can address multiple problem types, including but not limited to regression, classification, clustering, anomaly detection, and recommendation systems [39].

Evaluating the effectiveness of a machine learning model involves various metrics. In classification tasks, commonly used metrics include accuracy, precision, recall, F1-score, and ROC-AUC [40]. For regression tasks, mean absolute error, mean squared error, and R-square are commonly applied [38]. The choice of evaluation metric is contingent on the specific problem and objectives, making a thorough understanding of these elements vital for developing robust machine learning models.

3.2 Relevance of machine learning in environmental data analysis

Environmental systems are inherently complex, characterized by high variability and unpredictability. Traditional statistical models often struggle to capture the nuances of these systems due to their linear nature and the assumption of independence among predictors. Machine learning, on the other hand, offers the flexibility to model non-linear relationships and consider interaction effects among multiple variables, making it adept at capturing the complexity inherent in environmental systems [41].

Machine learning's strength lies not just in its ability to model complex systems, but also in its ability to handle unstructured data. Unlike traditional models that require structured tabular data, machine learning algorithms can handle a variety of data types, including text, images, and sound. For instance, images from remote sensing can be used in land use classification or change detection tasks using convolutional neural networks, a type of deep learning algorithm [42]. Text from social media posts can be analyzed using natural language processing techniques to gauge public sentiment towards environmental issues or to detect early signs of natural disasters [43].

Time-series analysis is another critical aspect of environmental data analysis. Environmental data is often collected over time, leading to a temporal sequence of observations. Machine learning plays a crucial role in the analysis of these time-series data. Recurrent neural networks (RNN) and Long Short Term Memory (LSTM) networks, types of deep learning algorithms, have shown significant promise in modeling temporal dependencies in environmental data, such as weather forecasting and prediction of air quality indices [44].

Lastly, the importance of feature engineering cannot be understated in the context of environmental data analysis. Feature engineering is the process of creating new features or modifying existing ones to improve model performance. While machine learning algorithms can automatically learn features in some cases (particularly in deep learning), domain knowledge can often guide more effective feature creation. For instance, in the case of predicting rainfall, features such as the season, geographical location, historical weather patterns, and other atmospheric conditions might be crucial. Incorporating such features in the model could potentially improve the model's predictive performance [38, 45].

3.3 Supervised, unsupervised, and reinforcement learning: a comparative analysis

This subsection aims to delve deeper into the nuances of supervised, unsupervised, and reinforcement learning, providing a comparative analysis across multiple dimensions.

Supervised Learning: Supervised learning algorithms learn from labeled training data to predict outcomes for unforeseen data [46]. The concepts of bias and variance are central to understanding the performance of these algorithms [47]. Bias refers to the error due to the model's assumptions in the learning algorithm, while variance pertains to the error from the model's sensitivity to fluctuations in the training set. An optimal balance between the two is crucial to avoid overfitting (high variance) and underfitting (high bias), both of which harm the model's predictive accuracy on new data. This balance is often achieved through methods such as cross-validation and hyperparameter tuning. Ensemble methods, including bootstrapping and bagging, and boosting methods, aim to enhance model performance by combining several weak learners to form a stronger overall model [38].

Unsupervised Learning: Unlike supervised learning, unsupervised learning algorithms discern patterns in data without pre-existing labels, primarily through clustering and dimensionality reduction techniques [48]. Concepts such as data density and similarity measures are instrumental in these techniques, and specific algorithms, such as hierarchical clustering and DBSCAN, utilize these concepts [49]. Handling outliers, determining the appropriate number of clusters (e.g., via the elbow method or silhouette score), are other critical aspects of unsupervised learning [50].

Reinforcement Learning: In reinforcement learning, an agent learns to perform actions in an environment to maximize a reward through the process of trial and error [51]. This learning paradigm is defined in terms of a Markov decision process, where the agent's actions in a certain state determine the probabilities of transitioning to other states [52]. The exploration versus exploitation dilemma is central to reinforcement learning, guiding the balance between trying new actions and exploiting known ones [53]. Techniques like Q-learning, SARSA, policy gradient methods, and more recent advances in deep reinforcement learning, provide frameworks to navigate this space [54].

Comparative Analysis: The three learning paradigms vary significantly in terms of data requirements, computational resources, interpretability, susceptibility to noise and outliers, and the level of human supervision required. For instance, supervised learning often requires large labeled datasets, which can be resource-intensive to collect, while unsupervised and reinforcement learning can work with unlabeled data. Supervised models are generally more interpretable than unsupervised and reinforcement learning models, making them suitable for applications where interpretability is a concern [55]. However, each of these learning paradigms has its strengths and weaknesses, and their appropriate usage largely depends on the specific problem at hand.

Understanding the subtleties of these paradigms is crucial for their effective application in environmental data analysis, given the complexity and high dimensionality of the data typically involved in this field [56].

4. Machine learning for enhanced IEQ assessment

In this section, we explore the practical applications of machine learning in assessing and enhancing Indoor Environmental Quality (IEQ) across a range of settings—residential, commercial, and specialized environments. We will examine how these algorithms contribute to optimized living conditions in homes, facilitate real-time environmental adjustments in commercial buildings, and ensure stringent IEQ compliance in specialized settings like healthcare facilities and data centers. Through these case studies, the section aims to provide an empirical substantiation of the machine learning methodologies discussed earlier, demonstrating their transformative potential in diverse IEQ contexts.

4.1 Residential buildings: Improving IEQ with machine learning

Machine learning has become a transformative force in improving Indoor Environmental Quality (IEQ) in residential buildings, offering significant advancements in Indoor Air Quality (IAQ), energy management, and occupancy detection.

Machine learning algorithms such as stochastic gradient boosting and support vector machines have demonstrated effectiveness in the domain of IAQ. These algorithms play a pivotal role in offering accurate predictions concerning room occupancy based on air quality indicators [57, 58]. Moreover, they are capable of efficiently classifying factors affecting IAQ under various conditions [59]. Other innovative approaches, including decision trees and hidden Markov models, have been applied to account for both real-time and anticipated occupancy states. These models have shown the capability to rectify sensor errors, thereby ensuring the reliability of IAQ parameters [60, 61].

Focusing on the challenges of sensor failures in pollutant monitoring, a significant study evaluated three Machine Learning (ML) algorithms – Multi-layer Perceptron (MLP), K-Nearest Neighbor (KNN), and Random Forest (RF) – for their effectiveness in classifying sensor readings as faulty or not. These algorithms demonstrated their superiority over standard statistical methods by creating better separation boundaries and utilizing contextual information. Numeric results from a 20-fold cross-validation displayed high average Area Under Curve (AUC) scores for each pollutant: 0.96 for MLP, 0.97 for KNN, and 0.97 for RF, indicating their robust performance in detecting faulty sensor readings [62].

Another key aspect of machine learning in IEQ is its application in energy efficiency. Studies have validated sensor data using techniques like artificial neural networks (ANNs) and Bayesian Networks, particularly focusing on U.S. commercial and residential buildings. Analyzing data from various sensors, these studies found that the root-mean-square error (RMSE) was generally less than 10% of the sensor's mean value for most sensors. However, the energy sensor data showed higher RMSEs, often exceeding 200%. Bayesian Networks, despite requiring longer training times, yielded lower errors compared to ANNs. Comparative RMSEs for ANNs and Bayesian Networks were 17.7% vs. 13.5% for liquid pressure, 5.0% vs. 2.1% for humidity, 1.9% vs. 1.1% for temperature, 0.03 vs. 0.02 for water flow, and 0.89 vs. 0.85 for energy consumption, highlighting the potential of these methods in enhancing sensor accuracy and energy efficiency [63].

In the field of residential energy management, machine learning, particularly Random Forest algorithms, has shown impressive efficacy in predicting home occupancy based on thermostat data. This was evident in a study evaluating various machine learning models, including heuristic baselines, traditional classifiers, and sequential models like recurrent neural networks (RNNs), for their ability to predict home occupancy. The Random Forest algorithm was noted for its high accuracy across different time horizons and efficient training for individual edge devices. Key numeric results included training times of 0.38 seconds for the Random Forest and 10 seconds for the RNN, with inference times of 0.001 seconds and 0.01 seconds, respectively, underscoring the potential of machine learning in adaptive thermal control in residential buildings [64, 65].

Further advancements in residential IEQ have been made using ensemble techniques and ANNs for the precise prediction of heating and cooling loads, especially in air-conditioned residential buildings in warm climates. One particular study focused on reducing energy consumption through energy-efficient building design, employing machine learning methods to estimate heating and cooling loads based on physical building characteristics. The performance of these methods was assessed using metrics like Mean Absolute Error (MAE) and Root Mean Square Error (RMSE), with significant improvements noted in cooling load prediction. The lowest MAE was 0.390, and the highest R^2 value was 0.996, indicating the model's high accuracy and efficiency [66, 67].

The integration of machine learning with smart home technologies has opened new avenues in IEQ management. For example, machine learning algorithms, used alongside IoT sensors, have enabled real-time monitoring and predictive analysis of indoor air pollutants and noise levels. Techniques such as fuzzy logic have been utilized to assess the efficacy of HVAC systems. These multifaceted applications of machine learning provide a robust foundation for enhancing IEQ in residential buildings and pave the way for a holistic approach to residential IEQ, marking a strong.

4.2 Commercial buildings: Real-time IEQ control with machine learning

Managing Indoor Environmental Quality (IEQ) in commercial buildings presents unique challenges due to varied occupancy patterns and the multifunctional nature of these spaces. The deployment of machine learning techniques has emerged as a key solution, offering real-time adaptability and predictive power for IEQ management.

Machine learning's versatility in real-time monitoring, classification, and predictive modeling is evident from studies that utilized algorithms like decision tree, random forest, and Support Vector Machine (SVM) for air quality and noise level predictions. Notably, SVM showed significant effectiveness in air quality prediction, with its accuracy improving from 75–95% after data quality adjustments. In contrast, decision tree and random forest algorithms achieved accuracies of 76% and 82%, respectively. For noise data classification, SVM achieved a remarkable 98% accuracy, consistently outperforming other models over a six-year evaluation period. The critical role of data quality and normalization in improving prediction accuracy was highlighted, with CO and SO2 sensors achieving high accuracies, while PM2.5 sensor accuracy was comparatively lower [68].

Despite these advancements, gaps in the literature are evident. Some studies have developed forecast models for CO2 buildup without elaborating on real-time IEQ monitoring [69]. Others have explored innovative areas like integrating machine learning with EEG signals for predicting IEQ conditions based on occupants' brainwave patterns [70]. Additional studies employed decision trees, hidden Markov models, and stochastic gradient boosting algorithms for immediate and future state occupancy prediction, demonstrating machine learning's potential for real-time IEQ control [57, 60]. The advantages of machine learning over traditional statistical approaches were also noted in IAQ sensor reading classification [62].

Specialized applications in commercial settings have seen the use of neural network modeling for identifying contaminant source positions and Sparse Spectrum Gaussian Process Regression (GPR) for real-time air quality prediction. These applications highlight machine learning's adaptability to various commercial environmental challenges [71, 72]. In educational and enterprise environments, studies have focused on ventilation rate control using IoT protocols and employing machine learning for air quality forecasting, further underscoring the diverse applicability of these technologies [73, 74].

Empirical evaluations often favor neural network-based models, using metrics such as RMSE, R2 score, and error rate. The ongoing advancements in these techniques, particularly in algorithm development and sensor calibration, promise enhanced precision and efficiency in monitoring indoor pollutants [75].

While machine learning offers promising opportunities for real-time IEQ control in commercial buildings, further research is needed for its effective implementation. Future research directions may explore the integration of machine learning with biometric signals for personalized environmental control [76].

4.3 Specialized environments: Strict IEQ regulation through machine learning

The challenge of maintaining optimal Indoor Environmental Quality (IEQ) in specialized environments, notably hospitals, has been rigorously researched. In such contexts, regulating IEQ is critical, impacting patient health, recovery, and satisfaction. Recognizing the limitations of conventional methods, researchers are increasingly turning to machine learning algorithms for more advanced computational approaches to these complexities.

Machine learning's application in hospitals has been pivotal in enhancing patient satisfaction, particularly concerning IEQ. A significant study at King Abdullah University Hospital (KAUH) in Jordan utilized Support Vector Machines (SVM) with a linear kernel and K-Nearest Neighbors (K-NN) to predict patient satisfaction. The SVM model, with a configuration of C = 0.01, proved more effective than other SVM kernels and K-NN. This effectiveness is evidenced by SVM's Pearson R-square value of 0.8948 with a P-value of .0001, surpassing K-NN's Pearson R-square of 0.2984. The study involved gathering self-reported data and field monitoring of environmental indicators within patient rooms at KAUH, using the same dataset for both training and testing. These outcomes highlight machine learning's role in providing insights that impact healthcare outcomes, particularly in enhancing IEQ to improve patient well-being [77].

Data collection in these environments often involves a mix of methodologies, including self-reported data, field monitoring, and sensor networks, creating diverse datasets for machine learning training and validation. For example, Elnaklah et al. combined sensor-generated data and human-reported satisfaction metrics to amass a comprehensive dataset [78].

Several machine learning models have shown effectiveness in regulating IEQ in hospital settings. An Autoregressive Hidden Markov Model (ARHMM) developed for a laboratory equipped with a sensor network demonstrated an average estimation accuracy of 80.78% in predicting occupancy patterns. This model, utilizing data from passive infra-red (PIR) sensors, CO2 concentration sensors, and relative humidity (RH) sensors, was benchmarked against classical Hidden Markov Models (HMM) and Support Vector Machines (SVM).

Bayesian networks have also been used to assess the risk of symptomatological complaints related to poor IEQ in Intensive Care Units (ICUs). A study in nine ICUs in João Pessoa, Brazil, analyzed temperature, noise, lighting, and air quality data, along with professional interviews. The Bayesian model revealed a 42.2% probability of physical symptoms and a 45.3% probability of psychological symptoms from environmental discomfort among ICU employees, with environmental temperature identified as a significant impacting factor [79, 80]. These probabilistic graphical models provide insightful perspectives on potential risk areas in IEQ within hospital settings.

Notably, there are gaps in existing literature, with several studies investigating IEQ through questionnaires and field studies without employing machine learning. For example, a study examining apartment and office buildings did not leverage machine learning, representing a missed opportunity for data-driven insights [81]. This highlights the significant potential for integrating machine learning techniques in future studies to provide more accurate, actionable recommendations for IEQ regulation in hospitals.

5. Conclusions

The chapter has provided an in-depth exploration of the integration of Machine Learning (ML) techniques into Indoor Environmental Quality (IEQ) assessment. IEQ, an aspect critical to human well-being, is intricately linked with numerous parameters including air quality, thermal, acoustical, and visual comfort. The chapter delineated the limitations of traditional methods in IEQ assessment, especially the lack of real-time data and predictive capabilities, and suggested machine learning as a potent solution for augmenting both the scope and accuracy of such evaluations. The implications are far-reaching, impacting residential buildings, commercial structures, and specialized environments such as healthcare facilities.

While machine learning appears to offer a robust approach to managing and enhancing IEQ, there are questions that remain unanswered. For instance, the ethical considerations of data collection and usage, particularly in specialized environments like healthcare facilities, warrant attention. Similarly, the interpretability of complex machine learning models and the implications for regulatory compliance are subjects requiring more in-depth study.

5.1 Future research directions

Data Ethics and Privacy: As machine learning increasingly intersects with IEQ assessment, ethical considerations around data collection and use will become more pressing. Future studies could focus on the ethical guidelines that ought to govern this intersection.

Interpretable Machine Learning Models: Future work could concentrate on developing more interpretable models without sacrificing predictive power, which is crucial for gaining regulatory approval and social acceptance.

Adaptive Systems: Research could delve into the development of adaptive machine learning systems that can evolve in real-time to meet the dynamic nature of IEQ elements, thereby offering more precise control mechanisms.

Personalization: Given the inter-individual differences in comfort and health responses, studies focusing on personalized IEQ assessment and control algorithms could offer a more nuanced approach to managing indoor environments.

Longitudinal Studies: To robustly assess the impact of machine learning on IEQ over time, future research could employ longitudinal study designs, perhaps integrated with natural experiments in real-world settings.

Cross-Sector Collaboration: As machine learning and IEQ assessment are topics with multidisciplinary ramifications, partnerships between AI researchers, environmental scientists, healthcare professionals, and policy-makers are essential for holistic solutions.

Resource Optimization: Machine learning algorithms capable of balancing both IEQ and energy efficiency would be a significant advancement, addressing the practical constraints of implementing ML in building management systems.

Global Standards: The development and adoption of global standards for machine learning in IEQ assessment could further facilitate interoperability and effectiveness across diverse building types and geographical locations.

Author details

Mohan Kumar Gajendran[1†], Ijaz Fazil Syed Ahmed Kabir[2†], Sudhakar Vadivelu[3], Eddie Yin-Kwee Ng[2*] and Ravi Chandra Thota[4]

1 Division of Energy, Matter and Systems, University of Missouri-Kansas City, Kansas City, MO, USA

2 School of Mechanical and Aerospace Engineering, Nanyang Technological University, Singapore

3 Department of Aerospace Engineering, Toronto Metropolitan University (TMU), Toronto, ON, Canada

4 Department of Computer Science, University of Missouri-Kansas City, Kansas City, MO, USA

*Address all correspondence to: mykng@ntu.edu.sg

† These authors contributed equally.

References

[1] Mujan I, Anelković AS, Munćan V, Kljajić M, Ružić D. Influence of indoor environmental quality on human health and productivity – A review. Journal of Cleaner Production. 2019;**217**:646-657 ISSN 0959-6526

[2] Ma N, Aviv D, Guo H, Braham WW. Measuring the right factors: A review of variables and models for thermal comfort and indoor air quality. Renewable and Sustainable Energy Reviews. 2021;**135**: 110436 ISSN 1364-0321

[3] The Atmospheric Fund. Improving Indoor Environmental Quality in Multi-Unit Residential Buildings. Toronto, ON: The Atmospheric Fund; 2019. Available from: https://taf.ca/custom/uploads/2019/05/Improving-Indoor-Environmental-Quality-in-Multi-Unit-Residential-Buildings-May-2019.pdf. [Accessed: December 23, 2023]

[4] Alhorr Y, Arif M, Katafygiotou M, Mazroei A, Kaushik A, Elsarrag E. Impact of indoor environmental quality on occupant well-being and comfort: A review of the literature. International Journal of Sustainable Built Environment. 2016;**5**(1):1-11 ISSN 2212-6090

[5] Mannan M, Al-Ghamdi SG. Indoor air quality in buildings: A comprehensive review on the factors influencing air pollution in residential and commercial structure. International Journal of Environmental Research and Public Health. 2021;**18**(6):3276. ISSN 1660-4601

[6] Kim H. Methodology for Rating a Building's Overall Performance Based on the ASHRAE/CIBSE/USGBC Performance Measurement Protocols for Commercial Buildings. Texas, USA: Texas A&M University; 2012

[7] Pedersen E, Borell J, Li Y, Stålne K. Good indoor environmental quality (ieq) and high energy efficiency in multifamily dwellings: How do tenants view the conditions needed to achieve both? Building and Environment. 2021; **191**:107581 ISSN 0360-1323

[8] Kishi R, Norbäck D, Araki A. Indoor Environmental Quality and Health Risk toward Healthier Environment for All. Vol. 1. Singapore: Springer;

[9] Persily A. Challenges in developing ventilation and indoor air quality standards: The story of ashrae standard 62. Building and Environment. 2015;**91**: 61-69 ISSN 0360-1323. Fifty Year Anniversary for Building and Environment

[10] Tien PW, Wei S, Darkwa J, Wood C, Calautit JK. Machine learning and deep learning methods for enhancing building energy efficiency and indoor environmental quality – A review. Energy and AI. 2022;**10**:100198 ISSN 2666-5468

[11] Fard ZQ, Zomorodian ZS, Korsavi SS. Application of machine learning in thermal comfort studies: A review of methods, performance and challenges. Energy and Buildings. 2022; **256**:111771 ISSN 0378-7788

[12] Mujeebu MA. Indoor Environmental Quality. London, UK, Rijeka: IntechOpen; 2019 ISBN 978-1-78985-252-3

[13] Laue J. Ashrae 62.1: Using the Ventilation Rate Procedure. 2018. Available from: https://www.csemag.com/articles/ashrae-62-1-using-the-ventilation-rate-procedure/ [Accessed: July 29, 2023]

[14] U.S. Environmental Protection Agency. Indoor Air Quality (iaq). Available from: https://www.epa.gov/indoor-air-quality-iaq [Accessed: July 29, 2023]

[15] World Health Organization. Environmental Health. Available from: https://www.who.int/europe/health-topics/environmental-health#tab=tab_1 [Accessed: July 29, 2023]

[16] Mujan I, Licina D, Kljajić M, Čulić A, Anelković AS. Development of indoor environmental quality index using a low-cost monitoring platform. Journal of Cleaner Production. 2021;**312**:127846 ISSN 0959-6526

[17] Kaushik A, Arif M, Ebohon OJ, Arsalan H, Rana MQ, Obi L. Effect of indoor environmental quality on visual comfort and productivity in office buildings. Journal of Engineering, Design and Technology, ahead-of-print(ahead-of-print). 2021;**21**:1746-1766

[18] Sarbu I, Sebarchievici C. Aspects of indoor environmental quality assessment in buildings. Energy and Buildings. 2013; **60**:410-419 ISSN 0378-7788

[19] Choi J-H, Loftness V, Aziz A. Post-occupancy evaluation of 20 office buildings as basis for future ieq standards and guidelines. Energy and Buildings. 2012;**46**:167-175 ISSN 0378-7788. Sustainable and healthy buildings

[20] Kim H et al. Field-test of the new ashrae/cibse/usgbc performance measurement protocols: Intermediate and advanced level indoor environmental quality protocols. ASHRAE Transactions. 2012;**118**:58

[21] Yao H, Cheng X, Shen W, Lv Y, Li A, Shen X. Sampling method for long-term monitoring of indoor environmental quality in residential buildings. Building and Environment. 2022;**215**:108965 ISSN 0360-1323

[22] Loomans MGLC, Mishra AK, Kooi L. Long-term monitoring for indoor climate assessment – The association between objective and subjective data. Building and Environment. 2020;**179**: 106978 ISSN 0360-1323.

[23] Roa CD, Schiavon S, Parkinson T. Targeted occupant surveys: A novel method to effectively relate occupant feedback with environmental conditions. Building and Environment. 2020;**184**: 107129 ISSN 0360-1323.

[24] Rasheed EO, Byrd H. Can self-evaluation measure the effect of ieq on productivity? A review of literature. Facilities. 2017;**35**(11/12):601-621

[25] Lai ACK, Mui KW, Wong LT, Law LY. An evaluation model for indoor environmental quality (ieq) acceptance in residential buildings. Energy and Buildings. 2009;**41**(9):930-936 ISSN 0378-7788

[26] Faqih F, Zayed T. Defect-based building condition assessment. Building and Environment. 2021;**191**:107575 ISSN 0360-1323

[27] Licina D, Wargocki P, Pyke C, Altomonte S. The future of IEQ in green building certifications. Buildings and Cities. London, United Kingdom: Ubiquity Press; 2021;**2**(1):907-927. DOI: 10.5334/bc.148

[28] Broday EE, Gameiro MC, da Silva. The role of internet of things (iot) in the assessment and communication of indoor environmental quality (ieq) in buildings: A review. Smart and Sustainable Built Environment. 2023; **12**(3):584-606

[29] Samuel AL. Some studies in machine learning using the game of checkers. IBM

Journal of Research and Development. 1959;**3**(3):210-229

[30] Rosenblatt F. The perceptron: A probabilistic model for information storage and organization in the brain. Psychological Review. 1958;**65**(6):386

[31] Nilsson NJ. Artificial intelligence prepares for 2001. AI Magazine. 1983; **4**(4):7-7

[32] Breiman L, Friedman J, Stone CJ, Olshen RA. Classification and Regression Trees. Oxfordshire, United Kingdom: Taylor & Francis;1984. pp. 1-358. ISBN: 9780412048418. Available from: https://books.google.com/books?id=JwQx-WOmSyQC

[33] Cortes C, Vapnik V. Support-vector networks. Machine learning. 1995;**20**: 273-297

[34] Hinton GE et al. Learning distributed representations of concepts. In: Proceedings of the Eighth Annual Conference of the Cognitive Science Society. Vol. volume 1. Amherst, MA: Taylor & Francis; 1986. p. 12

[35] Freund Y, Schapire RE. A decision-theoretic generalization of on-line learning and an application to boosting. Journal of Computer and System Sciences. 1997;**55**(1):119-139

[36] LeCun Y, Bengio Y, Hinton G. Deep learning. Nature. 2015;**521**(7553): 436-444

[37] Krizhevsky A, Sutskever I, Hinton GE. Imagenet classification with deep convolutional neural networks. Advances in Neural Information Processing Systems. 2012;**25**:84-90

[38] Gajendran MK, Rohowetz LJ, Koulen P, Mehdizadeh A. Novel machine-learning based framework using electroretinography data for the detection of early-stage glaucoma. Frontiers in Neuroscience. 2022;**16**: 869137. ISSN 1662-453X

[39] Gajendran MK, Khan MZ, Khan MA, Khattak. Ecg classification using deep transfer learning. In: In 2021 4th International Conference on Information and Computer Technologies (ICICT). Hawaii, USA: IEEE; 2021. pp. 1-5

[40] Khan MZ, Gajendran MK, Lee Y, Khan MA. Deep neural architectures for medical image semantic segmentation: Review. IEEE Access. 2021;**9**: 83002-83024

[41] Habibi S. Smart innovation systems for indoor environmental quality (ieq). Journal of Building Engineering. 2016;**8**: 1-13 ISSN 2352-7102

[42] Parkinson T, Parkinson A, de Dear R. Continuous ieq monitoring system: Context and development. Building and Environment. 2019;**149**: 15-25 ISSN 0360-1323

[43] Finch KC, Snook KR, Duke CH, King-Wa F, Tse ZTH, Adhikari A, et al. Public health implications of social media use during natural disasters, environmental disasters, and other environmental concerns. Natural Hazards. 2016;**83**(1):729-760

[44] Qi H, Xiao S, Shi R, Ward MP, Chen Y, Wei T, et al. Covid-19 transmission in mainland China is associated with temperature and humidity: A time-series analysis. Science of the Total Environment. 2020;**728**: 138778 ISSN 0048-9697

[45] Barandas M, Folgado D, Fernandes L, Santos S, Abreu M, Bota P, et al. Tsfel: Time series feature extraction library. SoftwareX. 2020;**11**:100456 ISSN 2352-7110

[46] Cunningham P, Cord M, Delany SJ. Supervised Learning. Berlin Heidelberg, Berlin, Heidelberg: Springer; 2008. pp. 21-49 ISBN 978-3-540-75171-7

[47] Belkin M, Hsu D, Ma S, Mandal S. Reconciling modern machine-learning practice and the classical bias–variance trade-off. Proceedings of the National Academy of Sciences. 2019;**116**(32): 15849-15854

[48] Ghahramani Z. Unsupervised Learning. Berlin Heidelberg, Berlin, Heidelberg: Springer; 2004. pp. 72-112 ISBN 978-3-540-28650-9.

[49] Saxena A, Prasad M, Gupta A, Bharill N, et al. A review of clustering techniques and developments. Neurocomputing. 2017;**267**:664-681 ISSN 0925-2312

[50] Askari S. Fuzzy c-means clustering algorithm for data with unequal cluster sizes and contaminated with noise and outliers: Review and development. Expert Systems with Applications. 2021; **165**:113856 ISSN 0957-4174

[51] Sutton RS, Barto AG. Reinforcement Learning: An Introduction. Massachusetts, London, England: MIT Press; 2018

[52] Puterman ML. Chapter 8 markov decision processes. In: Stochastic Models, volume 2 of Handbooks in Operations Research and Management Science. Amsterdam, Netherlands: Elsevier; 1990. pp. 331-434

[53] Jung D, Choi Y. Systematic review of machine learning applications in mining: Exploration, exploitation, and reclamation. Minerals. 2021;**11**(2):148. ISSN 2075-163X

[54] Gupta S, Singal G, Garg D. Deep reinforcement learning techniques in diversified domains: A survey. Archives of Computational Methods in Engineering. 2021;**28**(7):4715-4754

[55] Casert C, Vieijra T, Nys J, Ryckebusch J. Interpretable machine learning for inferring the phase boundaries in a nonequilibrium system. Physical Review E. 2019;**99**(2):023304

[56] Zhong S, Zhang K, Bagheri M, Burken JG, April G, Li B, et al. Machine learning: New ideas and tools in environmental science and engineering. Environmental Science & Technology. 2021;**55**(19):12741-12754

[57] Dela Cruz JC, Amado TM, Hermogino JQ, Andog MLC, Corpuz FT, Ng JRT, et al. Machine Learning-Based Indoor Air Quality Baseline Study of the Offices and Laboratories of the Northwest and Southwest Building of Mapúa University – Manila. In: 2020 11th IEEE Control and System Graduate Research Colloquium (ICSGRC). Shah Alam, Malaysia: IEEE; 2020

[58] Fritz H, Tang M, Kinney K, Nagy Z. Evaluating machine learning models to classify occupants' perceptions of their indoor environment and sleep quality from indoor air quality. Journal of the Air amp; Waste Management Association. 2022;**72**(12):1381-1397

[59] Saad SM, Shakaff AYM, Saad ARM, Yusof AM, Andrew AM, Zakaria A, et al. Analysis of feature selection with Probabilistic Neural Network (PNN) to classify sources influencing indoor air quality. In: AIP Conference Proceedings, Author(s). Penang, Malaysia: American Institute of Physics (AIP): 2017

[60] Ryu SH, Moon HJ. Development of an occupancy prediction model using indoor environmental data based on machine learning techniques. Building and Environment. 2016;**107**:1-9

[61] Deepa R, Raj KM, Balaji N, Durgadevi K. Machine learning based estimation of room occupancy using non-intrusive sensors. In: 2022 International Conference on Communication, Computing and Internet of Things (IC3IoT). Chennai, India: IEEE; 2022

[62] Seibert V, Araújo R, McElligott R. Sensor Validation for Indoor Air Quality using Machine Learning. In: Anais do Encontro Nacional de Inteligência Artificial e Computacional (ENIAC 2020). Pelotas, RS, Brazil: Sociedade Brasileira de Computação - SBC; 2020

[63] Smith MK, Castello CC, New JR. Machine learning techniques applied to sensor data correction in building technologies. In: 2013 12th International Conference on Machine Learning and Applications. Miami, FL, USA: IEEE; 2013

[64] Huchuk B, Sanner S, O'Brien W. Comparison of machine learning models for occupancy prediction in residential buildings using connected thermostat data. Building and Environment. 2019; **160**:106177

[65] Sadeghi A, Younes SR, Young WA, Weckman GR. An intelligent model to predict energy performances of residential buildings based on deep neural networks. Energies. Basel, Switzerland: MDPI; 2020;**13**(3):571. Availbale from: https://www.mdpi.com/1996-1073/13/3/571. ISSN: 1996-1073. DOI: 10.3390/en13030571

[66] Goyal M, Pandey M. A systematic analysis for energy performance predictions in residential buildings using ensemble learning. Arabian Journal for Science and Engineering. 2020;**46**(4): 3155-3168

[67] Asif S, Saba T. Machine learning for exploration of the factors affecting the building energy usage: A case study. Applied and Computational Engineering. 2023;**2**(1):694-699

[68] Shah SK, Tariq Z, Lee J, Lee Y. Real-time machine learning for air quality and environmental noise detection. In: In 2020 IEEE International Conference on Big Data (Big Data). Atlanta, GA, USA: IEEE; 2020

[69] Rastogi K, Lohani D. An Internet of Things Framework to Forecast Indoor Air Quality Using Machine Learning. Singapore: Springer; 2020. pp. 90-104

[70] Kim J, Kim H, Hong T. Automated classification of indoor environmental quality control using stacked ensembles based on electroencephalograms. Computer-Aided Civil and Infrastructure Engineering. 2020;**35**(5): 448-464

[71] Vukovic V, Tabares-Velasco PC, Srebric J. Real-time identification of indoor pollutant source positions based on neural network locator of contaminant sources and optimized sensor networks. Journal of the Air amp; Waste Management Association. 2010; **60**(9):1034-1048

[72] Arun G, Rathi S. Real Time Air Quality Evaluation Model using Machine Learning Approach. Journal of Information Technology and Digital World. Coimbatore, India: Inventive Research Organization; 2022;**4**(1):23-33. DOI: 10.36548/jitdw.2022.1.003

[73] Tagliabue LC, Cecconi FR, Rinaldi S, Ciribini ALC. Data driven indoor air quality prediction in educational facilities based on IoT network. Energy and Buildings. 2021;**236**:110782

[74] Iskandaryan D, Ramos F, Trilles S. Air quality prediction in smart cities

using machine learning technologies based on sensor data: A review. Applied Sciences. 2020;**10**(7):2401

[75] Saini J, Dutta M, Marques G. Indoor air quality prediction systems for smart environments: A systematic review. Journal of Ambient Intelligence and Smart Environments. 2020;**12**(5): 433-453

[76] Maaloul K, Brahim L. Comparative analysis of machine learning for predicting air quality in smart cities. WSEAS Transactions on Computers. 2022;**21**:248-256

[77] Wedyan M, Ali H, Abdullah M. Toward predicting Patient's satisfaction of indoor environmental quality in Jordanian hospitals using SVM and K-NN machine learning algorithms. In: 2022 13th International Conference on Information and Communication Systems (ICICS). Irbid, Jordan: IEEE; 2022

[78] Elnaklah R, Fosas D, Natarajan S. Indoor environment quality and work performance in "green" office buildings in the Middle East. Building Simulation. Berlin, Germany: Springer Nature; 2020; **13**(5):1043-1062. DOI: 10.1007/s12273-020-0695-1. ISSN: 1996-8744

[79] Han Z, Gao RX, Fan Z. Occupancy and indoor environment quality sensing for smart buildings. In: 2012 IEEE International Instrumentation and Measurement Technology Conference Proceedings. Graz, Austria: IEEE; 2012

[80] de Araújo Vieira EM, da Silva LB, de Souza EL. The influence of the workplace indoor environmental quality on the incidence of psychological and physical symptoms in intensive care units. Building and Environment. 2016; **109**:12-24

[81] Roulet C-A, Bluyssen PhM, Cox C, Foradini F. Relations between perceived indoor environment characteristics and well being of occupants at individual level. In: Healthy Buildings 2006. Lisbon, Portugal: LESO-PB, EPFL; 2006. Available from: http://infoscience.epfl.ch/record/78593

Chapter 6

Indoor Air Quality in Health Care Units (Case Study: Namazi Hospital, Shiraz, Iran)

Forough Farhadi, Saeid Chahardoli and Mehdi Khakzand

Abstract

Indoor air quality (IAQ) represents an important research focus due to its direct and substantial implications on human health outcomes. Existing research showed that substandard IAQ exacerbates the effects of airborne diseases. The objective of this chapter would be to explore the correlation among indoor air quality (IAQ), location of air outlet valves, and fluctuations in IAQ indicators within the cardiovascular care unit (CCU). In this regard, a combination of experimental and numerical methods has been utilized. These included direct IAQ measurements within the unit and the application of computational fluid dynamics to simulate indoor air conditions based on the collected experimental data. In this specific circumstance, the state of the air outflow valve and the condition of the air change rate significantly affect the enhancement of IAQ levels. To confirm this hypothesis, existing literature was thoroughly reviewed according to IAQ guidelines. In a similar vein, the study included measurements of emissions such as CO_2, CO, PM2.5, and PM10. Additionally, it examined the association relating to IAQ, air outlet placement, and dynamics of the emissions within the patient's room.

Keywords: indoor air quality, building performance, healthy indoor spaces, infectious diseases, computational fluid dynamics

1. Introduction

Indoor air quality (IAQ) in healthcare facilities has the greatest significance due to its enormous effects on patient well-being, staff productivity, and overall health outcomes [1]. Airborne diseases highlight the potential to enhance healthcare settings' preventative procedures [2, 3]. Comprehending and managing prevalent indoor pollutants can mitigate the potential health risks and adverse health consequences associated with indoor air contaminants [4, 5].

World Health Organization (WHO) estimates that air pollutants cause up to 7.3 million fatalities annually, of which 4.3 million are due to indoor air pollutants [6]. Recently, the Harvard School of Public Health conducted a study that noticed a correlation between mortality rates attributed to infectious diseases and the rising levels of particulate matter concentration [5, 7, 8]. The impact of IAQ is particularly consequential for these individuals as those with compromised immune systems could be exposed to potentially life-threatening infections within hospitals due to poor air quality [3, 9, 10].

Studies and efforts related to this field emphasize how poor IAQ magnifies the effects of airborne pathogens [11–13] and the enormous risk that respiratory infections carried by aerosols pose, particularly in compact, inadequately ventilated spaces [14]. Finding the causes of the rising transmission of airborne diseases and their mortality rate has drawn recent scientific attention. The damage caused to human health varies, and this issue depends on particle pollutants' concentration [6, 15, 16].

Preventing or managing airborne infectious illnesses would be improved by IAQ research, characterization, and increased interest in creating healthy indoor settings [17]. Standards have been established for assessing IAQ in a building for its intended purpose. As an example, the EPA has developed a series of specific reference procedures to precisely gauge the levels of individual pollutants [18–20], and the World Health Organization (WHO) set a comparison of different indoor air quality guidelines [21–23]. These methods provide outstanding accuracy and precise time measurements, yet they present challenges such as the requirement for quality control assessments, frequent calibration, significant costs, and the necessity for an operator possessing specialized expertise [24–26]. Given the damaging impact on IAQ and individuals' vulnerability within healthcare facilities, addressing this concern in treatment areas holds paramount significance.

Recently, computational fluid dynamics (CFD) has potential techniques for analyzing particles' behavior in a room [27, 28]. It is influenced by a number of variables, including airborne contaminants, ventilation systems, and building layout [29, 30]. To maintain a balanced airflow and avoid recirculating contaminated air, the placement of air outflow valves is essential [31].

This issue should be considered in hospitals in order to control and monitor the microbiological quality of indoor air [32, 33]. At present, there exists a dearth of understanding among individuals concerning the assessment, recognition, and possible health consequences of indoor air quality (IAQ) [34–36]. Hence, the ongoing monitoring and regulation of indoor air quality within hospitals are integral components of infection prevention strategies and the promotion of a healthful indoor environment.

Recent studies investigated the importance of IAQ monitoring according to guidelines [31, 36, 37]. Abdel-Salam et al. investigated PM10, PM2.5, and CO_2 experimentally for 24 h in urban homes in Egypt based on WHO guidelines [21]. Kephart et al. investigated NO_2 measurements for 48 h in Homes in Peru based on WHO guidelines [38]. Woolley et al. studied CO concentration from Wednesday to Friday evenings for 48 h in home residence apartments based on WHO guidelines in the United Kingdom [39]. Amadeo et al. studied O_3, NO_2, SO_2, and PM10 measurements from December 2008 to December 2009 in schools based on WHO guidelines in Guadeloupe (French West Indies) [40]. Shen et al., Huang et al., and Abdel-Salam investigated IAQ measurements based on EPA guidelines [41–43]. Poulin et al. studied IAQ measurements based on Canadian IAQ guidelines [44].

Also, Singh et al. investigated PM2.5 for two weeks of measurements in a commercial shopping complex (CSC) based on NIOSH guidelines in Delhi.

Branco and colleagues conducted a study on indoor air quality (IAQ) and discovered that children sensitized to common aeroallergens had an increased likelihood of developing childhood asthma when exposed to particulate matter [45]. These factors can have adverse effects on human well-being, potentially causing disruptions to daily life [25, 46–48]. Connections have been demonstrated to exist between particular indoor exposures, even when they are at low levels. This can pose health risks, particularly for individuals who have not developed prior sensitivities [49–52].

Groulx et al. have concluded that alterations in outdoor air quality significantly affect the composition of medical air administered to patients. They also emphasize the crucial significance of monitoring and controlling the quality of medical air within healthcare facilities [53]. In their study, Jung et al. examined levels related to significant airborne pollutants, which encompassed CO, CO_2, O_3, total volatile organic compounds (TVOC), formaldehyde HCHO, and particulate matter (PM2.5 and PM10). Their research revealed that inpatient rooms had significantly elevated levels of CO_2 and TVOC compared to nursing stations, clinics, and clinic waiting rooms [54]. In contrast, Nair et al. [55] verified that pollutants including particulate matter (PM10, PM2.5), NO_2, SO_2, CO, O_3, and CO_2 elevate the risk of contracting airborne illnesses, resulting in prolonged infectiousness of airborne viruses and contributing to an unhealthy environment.

Recent studies used short-period monitoring as a technique to analyze the quality of indoor air [21, 38, 56]. Piexoto et al. [57] experimentally investigated CO_2 and CO in the fitness center. Related activities in healthcare facilities induce specific emissions. This condition may become harmful if it exceeds the acceptable limits and may accelerate the virus's contagion [58–60]. As a result, the particles transferred with the incoming air would affect the transmission of infectious diseases. Recent studies based on IAQ measurements show that this experimental system is an effective method to help us understand the status of the environment and can prevent the chances of infectious transition and mechanically or naturally reach the appropriate IAQ. This is effective because it can lead to more healthy spaces, especially in healthcare facilities. Within this context, the crucial role of maintaining indoor air quality in infection control becomes evident. Inspired by the above-mentioned facts, the objective of this study is to contribute scientific insights into indoor air quality (IAQ) within healthcare facilities. This will be achieved through experimental measurements of IAQ indicators, which will then be compared to the acceptable limits established by IAQ guidelines. Research focused on health underscores the growing importance of comprehending the interplay between the built environment and the transmission of infections [61–63]. Examining the dispersion and mobility patterns of indoor particles can improve indoor air quality and promote the sustainable and healthy development of indoor environments. Recently, government agencies have implemented regulations and enforcement measures to enhance environmental health by controlling outdoor air pollutants. Different guidelines have been established to help monitor the air quality both indoors and outdoors [64].

Since 1979, WHO has consistently addressed indoor environmental conditions in numerous reports [65], with the objective of ensuring adequate indoor air quality, particularly within hospital facilities [66, 67].

According to WHO, numerous indoor air pollutants can adversely impact both the indoor environment and human health [68]. Airborne pollutants, including VOCs, PMs, SO_2, CO, NO, PAHs, microbial spores, pollen, allergens, and more, are the primary contributors to the deterioration of IAQ [69]. In their study, Kim et al. have come to the conclusion that high levels of PM10 have the potential to transmit indoor infections in closed spaces [36].

As awareness of the significance of indoor environments for human health grew, scientists began proposing various recommendations. These recommendations include establishing optimal air exchange rates within specific timeframes, regulating the subsequent emission of air pollutants from various products, and establishing a foundation of guidelines and references for indoor environmental considerations [70]. Achieving an ideal air changes per hour (ACH) is crucial in IAQ within healthcare facilities to maintain efficient ventilation and reduce the danger of airborne pollution.

In previous research, it has been established that low air changes per hour (ACH) and insufficient ventilation negatively impact occupants' health. Recent analyses have revealed that ACH rates in many European countries range from approximately 0.35–1 ACH [71], while in China, they vary from about 0.35 to 0.78 ACH [72]. These earlier studies predominantly concentrated on aspects like achieving net zero energy buildings (NZEB), optimizing thermal comfort, and reducing energy consumption, which often led to lower ACH rates. However, a significant knowledge gap exists regarding the design criteria. This gap relates to determining the ideal ACH thresholds and achieving the best air quality with the lowest health risk for occupants.

In this chapter, we have conducted numerical simulations to replicate fluid dynamics using the RNG k-e turbulence model and to analyze the motion of particles using the discrete particle model (DPM). These simulations are employed to investigate the behavior of particles within the unit and their interaction with the surrounding fluid. The study's findings should provide an extensive understanding of the ventilation system design by air outlet valve height on particle dispersion and removal in situations when high outdoor contamination loads are an issue.

This chapter is based on CO, CO_2, PM2.5, and PM10 measurements in a patient's room in the CCU of Namazi Hospital in Shiraz, Iran. Based on previous research, IAQ measurements have been compared to IAQ guidelines (EPA, NIOSH, WHO, and Canadian). This study aims to understand the air quality inside Namazi Hospital, particularly in the Cardiovascular Care Unit. We're looking at how the direction of air outlet valves affects the indoor air quality and its potential impact on preventing the spread of diseases.

The study follows this sequence: (i) Experimental measurements are carried out, (ii) the obtained results are compared to IAQ guidelines, and (iii) the findings from this investigation are analyzed utilizing CFD computational design.

2. Method

This study is based on experimental data collection, CFD modeling, and analysis, commencing with a real-world case study. Namazi Hospital has been chosen for a case study due to its location in a congested area of Shiraz, Southern Iran, with a dry and warm climate. Measurements were undertaken both indoors and outdoors within the CCU (critical care unit). The sampling encompassed pollutants like CO, CO_2, PM2.5 and PM10. Monitoring was carried out during typical daily activities and under conditions representative of occupancy, following the ISO 4224 standard. To simulate the patient's breathing zone accurately, the measurement devices were positioned on a level surface at a 1-m height, maintaining a minimum distance of 1 m from any doors or active heating equipment. The devices used were calibrated before and after measurements.

2.1 IAQ sampling and analysis

The sampling devices were securely positioned in locations equipped with electricity, shielded from direct sunlight and rain within a protective shelter. They were positioned at a height of 1 m above the ground to ensure unobstructed access for the sampling inlets and sensors, thus preserving the integrity of the sampling process. The levels of CO, CO_2, PM2.5, and PM10 were simultaneously measured alongside other air parameters both indoors and outdoors at 10:20 AM and 6:20 PM on Monday, March 1, 2021. The following measurements were conducted using the following equipments: IAQ for CO (Aeroqual, model S200, N/A), AQ100 for CO_2 (Aeroqual, model S200, 250313-2307), DustTrak for PM2.5 and PM10 (model ISI, 21221), and a flow meter (model N/A) (**Figure 1**).

Measurements have taken place in Room 6, which has an area of 14.5 m^2 and occupants on every shift. This room is mechanically ventilated by 2 inlets with a velocity of 2.5 m/s, dimensions of 25×25 cm, and an outlet that has a dimension of 50×50 cm. Windows and doors were closed during the measurement period. The results have been compared to IAQ guidelines (EPA, NIOSH, WHO, LEED, and Canadian).

This study centers on the utilization of computational fluid dynamics (CFD) and particle tracking to model indoor air quality (IAQ) within healthcare environments. In this context, the Reynolds-averaged Navier-Stokes (RANS) equations serve as the foundation for describing turbulent, incompressible airflow. To address various turbulence scales effectively, a customized RNG k-ε model is employed.

In this research, both Eulerian and Lagrangian methods are simultaneously used to model indoor airflow and particle trajectories. The particle motion equation includes gravity and drag forces. The accuracy of particles' simulation is ensured by integrating RANS models with the model of discrete random walk (DRW).

Regarding the deposition of particles, wall-normal turbulent velocity fluctuations play a crucial role. Particle deposition on surfaces is assumed when particles hit the walls.

For computational simulations, ANSYS-FLUENT 18.2 software is utilized, with an Intel(R) core (TM) i7-6800 CPU @ 3.40 GHz processor and 32 GB RAM. The airflow conditions include a velocity of 2.5 m/s and air density of 1.225 kg/m^3. Particle sizes range from 1 to 10 μm, with a density of 2000 kg/m^3. A one-way method is employed due to diluting particle concentration.

The study also examines particle deposition on walls in a patient room using a Eulerian-Lagrangian approach and validates results against other literature. Various pollutants are monitored in this mechanically ventilated room.

Figure 2 illustrates how the velocity magnitude's competitors vary across diverse computational grids when employing the realizable k-ε model. Grid independence

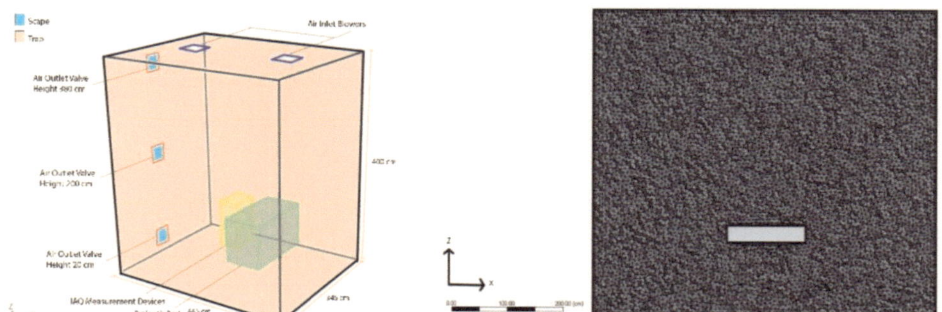

Figure 1.
Created the room geometry description drawing and the fine unstructured mesh.

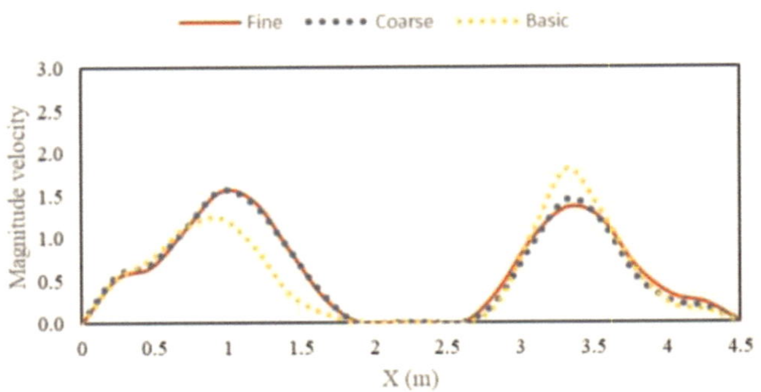

Figure 2.
Grid independence for velocity magnitude with 100; 500,000; and 1,000,000 grids.

shows the domain with 1 million mesh has the best accuracy for fluid flow simulation. The velocity profiles have been acquired and are being compared across varying grid resolutions at heights of 1, 2, and 3 m. Additionally, for the solving of the governing equation, the SIMPLE algorithm, as well as the finite volume method, are employed.

2.2 Mathematical equation and model description

The Reynolds-averaged Navier-Stokes (RANS) equations are applied to describe turbulent airflow that remains incompressible. Within this context, we observe the utilization of the continuity and momentum equations [73]:

$$\frac{\partial \overline{U}_i}{\partial x_i} = 0, \tag{1}$$

$$\rho \overline{U}_j \frac{\partial \overline{U}_i}{\partial x_j} = -\frac{\partial \overline{P}}{\partial x_i} + \frac{\partial}{\partial x_i}\left[\mu\left(\frac{\partial \overline{U}_i}{\partial x_i} + \frac{\partial \overline{U}_j}{\partial x_i}\right) - \rho\overline{u_i u_j}\right], \tag{2}$$

Here, $\overline{U}_i$ stands for the fluctuation velocity, and $\overline{U}$ and $\overline{P}$ stand for the average speed and pressure, respectively. In addition, $(\overline{-u_i u_j})$ and μ are the fluid viscosity and the Reynolds stress tensor, additionally, ρ is density.

Yakhot et al. [73] developed the RNG k-model by applying a process of renormalization to the Navier-Stokes equations, taking into account diverse scales of turbulent motion. RNG k-model's transport equations are provided as:

$$\frac{\partial}{\partial x_i}(\rho k u_i) = G_k - Y_k + \frac{\partial}{\partial x_i}\left[\left(\mu + \frac{\mu_t}{\sigma_k}\right)\frac{\partial k}{\partial x_j}\right], \tag{3}$$

$$\frac{\partial}{\partial x_i}(\rho \varepsilon u_i) = C_{\varepsilon 1}\frac{\varepsilon}{k}G_k - C_{\varepsilon 2,RNG}\rho\frac{\varepsilon^2}{K} + \frac{\partial}{\partial x_i}\left[\left(\mu + \frac{\mu_t}{\sigma_k}\right)\frac{\partial k}{\partial x_j}\right], \tag{4}$$

here:

$$Y_k = \rho\varepsilon \ and, S = \sqrt{2S_{ij}S_{ij}}, G_k = S^2\mu_t, S_{ij} = \frac{1}{2}\left(\frac{\partial u_i}{\partial x_j} + \frac{\partial u_j}{\partial x_i}\right), \tag{5}$$

$$C_{\varepsilon 2,RNG} = C_{\varepsilon 2} + \frac{C_\mu \eta^3 \left(1 - \frac{\eta}{\eta_0}\right)}{1 + \beta \eta^3} \eta = \left(2S_{ij}S_{ij}\right)^{\frac{1}{2}} \frac{k}{\varepsilon}, \eta_0 = 4.337, \beta = 0.012, \tag{6}$$

The importance of demonstrating particle dispersion inside buildings has led to specific concentrations being determined for simulating diffusion and deposition. Lagrangian particle trajectories offer advantages as they enable the tracking of random particles within the computational domain while also facilitating a straightforward interaction between the particles and walls, thanks to well-defined boundary conditions. In the present study, both Eulerian and Lagrangian methods are employed to simultaneously compute the indoor flow field and particle trajectories. Gravity and drag force are included in the particle motion equation as follows:

$$\frac{du_i^p}{d_t} = \frac{C_D Re_p}{24\tau} \left(u_i - u_i^p\right) + g_i, \tag{7}$$

Combining the particle velocities yields the particle position as follows:

$$\frac{dx_i}{d_t} = u_i^p, \tag{8}$$

RANS models are used to calculate the kinetic energy of turbulence, mean flow velocity fields, and turbulence dissipation rate. They combine realistic simulation of particle motion with precise modeling of fluid flow fluctuations. The following equations [74] are used in the DRW model for this purpose:

$$u' = \zeta u'_{rms}, v' = \zeta v'_{rms}, w' = w'_{rms} \tag{9}$$

Ref. [74] Research has indicated that achieving precise forecasts for the velocity of sediment particles moving through a duct necessitates the consideration of turbulent velocity fluctuations in the wall-normal direction within the near-wall region. Lecrivain et al. [75] wrote the equation [11]. To accurately model particle deposition, complicated wall turbulent flow is represented as follows:

$$\sqrt{\overline{u_2'^2}} = u^* \left(\frac{a_1 y^{+2}}{1 + b_1 y^+ + c_1 y^{+2.41}}\right), \quad for\, y^+ < 30 \tag{10}$$

here, c_1 = 0.0014, b1 = 0.203, and a_1 = 0.0116 y plus can be described by,

$$y^+ = \frac{yu^*}{v} \tag{11}$$

3. Result

This study relies on the collection of experimental data at Namazi Hospital. The findings magnified the importance of keeping levels of CO, CO_2, PM2.5, and PM10 concentrations low for various health-related purposes, possibly even in transmission of SARS-CoV-2. Results showed the measurements of CO and CO_2 concentrations have been lower than acceptable limitations offered by IAQ guidelines (EPA, NIOSH,

WHO, LEED, and Canadian). Also, indoor and outdoor measurements of PM2.5 and PM10 have been above LEED and EPA guidelines.

Utilizing CFD as an analytical tool, its influence on indoor air quality and ventilation has been investigated and specifically applied to facilitate particle dispersion in an intensive care unit (ICU), with a primary research objective of assessing how variations in outlet valve heights impact indoor air quality. In space, on Monday, March 1, 2021, at 10:20 a.m. and 18:20 p.m., we measured the pollutants CO, CO_2, PM2.5, and PM10 in the room at two separate times of day.

For the simulation phase, we employed ANSYS Fluent to model the room based on its actual dimensions. The inlet velocity in this room was 2.5 m/s. The entrance valves measure 25 by 25 cm, while the output valves measure 50 by 50 cm. The grid independence computation determined that an unstructured mesh with roughly 1,106 cells should be installed. Then, to replicate airflow in the room, we used RNG k. Based on the outcomes of the simulation, we found that the airflow velocity was higher when the exit valve was placed 20 cm from the wall.

The domain was then filled with particles that were 1, 2.5, 4, 7.5, and 10 mm in diameter. By altering the location of the outlet valve, we successfully reduced contamination in the room equipped with an airflow ventilation system. Different air outlet valve heights have an impact on the flow field and particle deposition. This issue has been investigated by using velocity contours and streamlines. The outcome shows that a higher exit valve would result in particle entrapment in the space. As the height of the outlet valve decreases, the flow of air would increase, and the particles' concentration would be removed. Therefore, the chances of infectious transmission will be reduced.

The calculated ACH for the case studies, based on a room area of 61.41 m^3 (3.45 m $\times$ 4.45 m $\times$ 4 m), two inflows totaling 0.062 m^2, an inlet velocity of 5.5 m/s, and an airflow rate of 0.15 m^3/s, falls within the acceptable IAQ range at 8.8 ACH (**Figure 3**).

Based on the simulation results, the inlet valve is strategically situated at a height of 20 cm, accounting for the natural tendency of cold air to accumulate at lower levels and facilitating its exit through the lower outlet valve. Additionally, the higher airflow velocity, as represented by the yellow color in **Table 1**, signifies efficient air movement. This observation, coupled with air change rates meeting industry standards, promises improved indoor air quality.

Figure 4 show the dispersion of the particles inside the room and illustrate how convenient it is for particles to exit the room if they are placed at 20-cm height.

DPM concentration at the 20-cm outlet height varies with time. Initial seconds exhibit high emissions near the outlet, but after 90 s, CFD indicates substantial particle removal from the room. This indicates effective ventilation dynamics.

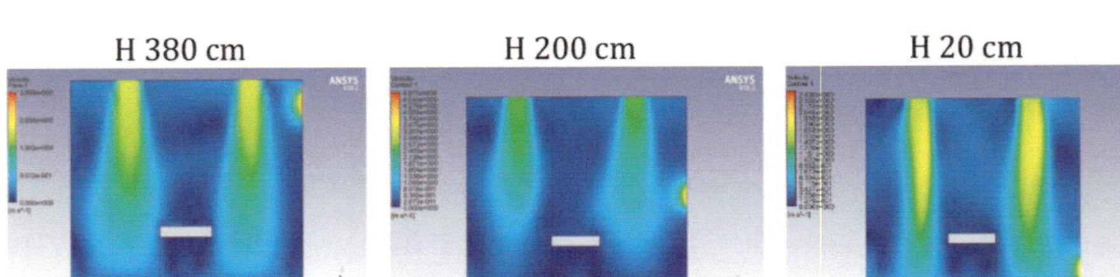

Figure 3.
A contour velocity magnitude for the model at the z/x plane located at the center surface.

DOI: http://dx.doi.org/10.5772/intechopen.113724

Date (Time)	Indoor				Outdoor			
	PM2.5 (Mg/m^3)	PM10 (Mg/m^3)	CO (ppm)	CO_2 (ppm)	PM2.5 (Mg/m^3)	PM10 (Mg/m^3)	CO (ppm)	CO_2 (ppm)
10:20	0.005	0.011	1.16	913.0	0.017	0.023	4.03	668.0
18:20	0.005	0.012	0.16	796.0	0.006	0.015	5.17	774.0

Table 1.
Indoor and outdoor measurements [31].

4. Discussion

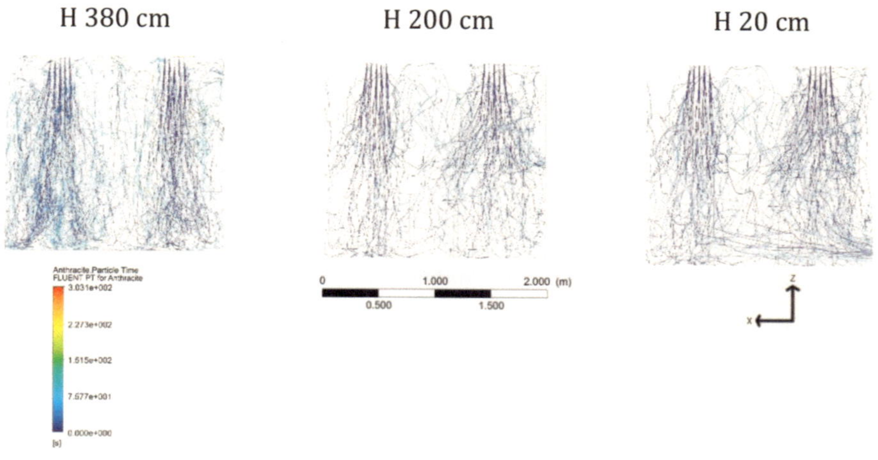

Figure 4.
Particle dispersion inside the room.

The comfort, productivity, and health of building inhabitants are directly impacted by indoor air quality. Indoor pollution, which has been linked to 4.1% of global fatalities in recent decades, can have immediate or long-term health consequences. Recent studies utilizing indoor air quality (IAQ) and computational fluid dynamics (CFD) demonstrate how useful this experimental approach is for assisting architects in producing improved designs. This method provides crucial information about airflow patterns and their effect on pollutant particles, enhancing architects' understanding of IAQ. This approach is effective as it bridges the gap between technological analysis, such as the CFD results, and architectural design.

A good indoor air quality is important in operating rooms where contaminated air can cause surgical site infections. An appropriate ventilation strategy, underscored by a comprehensive understanding of airflow patterns and pollutant dispersion, is essential for controlling and reducing indoor pollution while optimizing the performance of the ventilation system. This study employed computational fluid dynamics to analyze airflow patterns and the spread of airborne contaminants within indoor settings.

In recent research, there has been an emphasis on the surveillance of indoor air quality [27, 76, 77]. Ascione et al. [78], in a study on university classrooms in Italy, by investigation of IAQ measurements proposed new scenarios and provided supporting evidence regarding the effectiveness of the systems responsible for thermal comfort that wouldn't pollute airflow. Additionally, they assessed the appropriateness of certain air distribution strategies, such as ceiling squared and linear slot diffusers, in comparison to conventional methods. Leconte et al. [79] conducted an experimental assessment of airflow and inventive active air ducts impact IAQ in a residential building. Cetin at al. [80] examined efficiency of varying air exchange rates on the dispersion and settling of indoor particles within a ventilated room.

Dobson et al. [81] investigated the quality of indoor air based on WHO guideline limits, in residential homes in Scotland, Florence, Greece, Milan, and Catalonia. The findings indicated that, in the context of this study, only a small number of households achieved complete smoke-free status. Lewis et al. [82], by investigating PM2.5 measurements from December 2011 to January 2012 in 105 residential homes in India, found out that in houses in which traditional stoves were used, pollution levels still remained above WHO guidelines.

Can et al. [83] studied NO_2, O_3, and VOCs for 7 days in a university in Turkey, the lifetime cancer risks for individuals employed within the department, including faculty members and technicians, exceeded the acceptable risk threshold established by the USEPA. Shen et al. [41] surveyed the IAQ within healthcare facilities, and based on the research findings, proposed the potential use of AgZ filtering as a means to manage bacteria and fungi parameters in hospital settings for indoor air quality management.

Baurès et al. [84] explored the levels of chemical and microbiological compounds present in the indoor air of two hospitals in France, which found that these concentrations (aldehydes, limonene, phthalates, aromatic hydrocarbons), exist in the space even low and are related to ventilation efficiency. Baboli et al. [85], conducted an investigation into the airborne transmission of infectious diseases at Razi Hospital in Iran, specifically chosen for this study. The findings corroborate the presence of airborne transmission of SARS-CoV-2 bioaerosols indoors. On May 7, 2020, Kenarkoohi et al. [7] conducted research into the transmission of the COVID-19 virus among confirmed COVID-19 patients within the indoor air of hospital wards.

These investigations clearly showed a strong influence on IAQ monitoring and controlling to achieve healthy spaces. There has been limited research conducted on indoor air quality (IAQ) within healthcare facilities. Furthermore, certain scholars propose exploring strategies for regulating and reducing pollution levels. The aim of this study is to assess the indoor air quality in Namazi Hospital's critical care unit (CCU) and to confirm whether the room is healthy or not for patients, also, reduce the possibility of the transition of infectious diseases by non-polluted air.

We came to the fact that the indoor and outdoor measurements of PM2.5 and PM10 have been higher than the guidelines' limitations. This condition can make it easier for infectious diseases to spread, especially those that are transmitted through contagious means. This matter has the potential to be for the corrupted HVAC filters and misfunctioning outlet valves or HVAC systems. Indoor and outdoor air should be monitored frequently for early detection of possible ventilation problems. This investigation suggests that frequent IAQ monitoring can lead to healthier spaces.

The ACH (air changes per hour) in numerous European countries for residential buildings ranges from approximately 0.35–1 [71]. However, the recommended protocol to minimize airborne infection transmission, especially during the COVID-19 pandemic, suggests a minimum of 12 ACH [86].

5. Conclusion

Recent studies showed the impact of IAQ on the transmission of infectious diseases. Also, how monitoring and controlling IAQ can be effective. Therefore, the IAQ of the CCU of Namazi Hospital was studied both experimentally and conceptually to understand the indoor air quality and its consequences on outlet valve height. It was found that the IAQ with respect to CO, CO_2, PM2.5, and PM10 was critical in the patient room due to room procedure's significance and the count of individuals present. If the IAQ indicators were above the guidelines' limitations, the chances of catching airborne diseases would increase sharply. In conclusion, frequently monitoring indoor air will be more effective in making a better IAQ and reducing the likelihood of infectious diseases transmission becomes more probable. Secondly, an observation revealed measurements of PM2.5 and PM10 within the patient's room in the critical care unit (CCU) in Namazi Hospital were above IAQ guidelines limitations.

Based on ANSYS Fluent output, particle concentration remains trapped when the outlet is at 380 cm due to airflow pressure. At 200 cm, emissions escape, while at 20 cm, substantial airflow removes particles. The highest deposit fraction occurs at 380 cm, resulting in the lowest air quality. Conversely, placing the valve at the bottom yields better particle removal.

We came to the fact that if the outlet valve position is above the ground height, mechanical ventilation systems will not provide enough air renewal throughout the patient room's interior space. And this would lead to the trapped and built up condition of the air pollutants, especially in highly polluted locations. Therefore, the IAQ will be lowered, and the likelihood of spreading infectious diseases will increase. It is advised to use these principles since lowering the particle deposition fraction is needed more than other options. Setting the exit valves lower can significantly alter the air in the space and help with incoming flow egress. By managing this matter, we could enhance our effectiveness in reducing the probability of airborne disease transmission within vulnerable environments such as healthcare facilities. There were few studies with interventions to improve IAQ, IAQ monitoring in healthcare spaces, and investigating the impact of IAQ on health based on IAQ guidelines.

Acknowledgements

We would like to express our gratitude to Namazi Hospital, particularly the Occupational Health Department, for their invaluable cooperation and guidance that significantly contributed to this research.

Conflict of interest

The authors declare no conflict of interest.

Author details

Forough Farhadi, Saeid Chahardoli and Mehdi Khakzand*
School of Architecture and Urban Design, Iran University of Science and Technology, Tehran, Iran

*Address all correspondence to: mkhakzand@iust.ac.ir

References

[1] Facciolà A, Pellicanò GF, Visalli G, et al. The role of the hospital environment in the healthcare-associated infections: A general review of the literature. European Review for Medical and Pharmacological Sciences. 2019;**23**:1266-1278

[2] Alvarenga MOP, Dias JMM, Lima BJLA, et al. The implementation of portable air-cleaning technologies in healthcare settings – A scoping review. The Journal of Hospital Infection. 2023; **132**:93-103

[3] Fonseca A, Abreu I, Guerreiro MJ, et al. Indoor air quality in healthcare Units&mdash: A systematic literature review focusing recent research. Sustainability. 2022;**14**. Epub ahead of print 2022. DOI: 10.3390/su14020967

[4] Deng S, Lau J. Seasonal variations of indoor air quality and thermal conditions and their correlations in 220 classrooms in the Midwestern United States. Building and Environment. 2019;**157**: 79-88

[5] Wu X, Nethery RC, Sabath MB, et al. Air pollution and COVID-19 mortality in the United States: Strengths and limitations of an ecological regression analysis. Scientific Advance. 2020;**6**. Epub ahead of print November 2020. DOI: 10.1126/sciadv.abd4049

[6] Correia G, Rodrigues L, Gameiro da Silva M, et al. Airborne route and bad use of ventilation systems as non-negligible factors in SARS-CoV-2 transmission. Medical Hypotheses. 2020; **141**:109781

[7] Kenarkoohi A, Noorimotlagh Z, Falahi S, et al. Hospital indoor air quality monitoring for the detection of SARS-CoV-2 (COVID-19) virus. Science Total Environment. 2020;**748**:141324

[8] Domingo JL, Rovira J. Effects of air pollutants on the transmission and severity of respiratory viral infections. Environmental Research. 2020;**187**: 109650

[9] Ghinai I, McPherson TD, Hunter JC, et al. First known person-to-person transmission of severe acute respiratory syndrome coronavirus 2 (SARS-CoV-2) in the USA. Lancet (London, England). 2020;**395**:1137-1144

[10] Khakzand M, Chahardoli S. Indoor thermal comfort and high residential buildings (Case study: Two highrise buildings in two different microclimates in Tehran city). Journal of Urban Management Energy Sustainable. 2023;**3**: 124-142

[11] Conticini E, Frediani B, Caro D. Can atmospheric pollution be considered a co-factor in extremely high level of SARS-CoV-2 lethality in Northern Italy? Environmental Pollution. 2020;**261**: 114465

[12] Setti L, Passarini F, De Gennaro G, et al. SARS-Cov-2RNA found on particulate matter of Bergamo in Northern Italy: First evidence. Environmental Research. 2020;**188**: 109754

[13] van Doremalen N, Bushmaker T, Morris D, et al. Aerosol and surface stability of HCoV-19 (SARS-CoV-2) compared to SARS-CoV-1. medRxiv : The preprint server for health sciences. 2020. Epub ahead of print 2020. DOI: 10.1101/2020.03.09.20033217

[14] Klompas M, Baker MA, Rhee C. Airborne transmission of SARS-CoV-2: Theoretical considerations and available evidence. JAMA. 2020;**324**:441-442

[15] Toyinbo O, Phipatanakul W, Shaughnessy R, et al. Building and indoor environmental quality assessment of Nigerian primary schools: A pilot study. Indoor Air. 2019;**29**:510-520

[16] Reddy M, Heidarinejad M, Stephens B, et al. Adequate indoor air quality in nursing homes: An unmet medical need. Sciene Total Environment. 2021;**765**: 144273

[17] Pourkiaei M, Romain A-C. Scoping review of indoor air quality indexes: Characterization and applications. Journal of Building Engineering. 2023;**75**: 106703

[18] Piedrahita R, Xiang Y, Masson N, et al. The next generation of low-cost personal air quality sensors for quantitative exposure monitoring. Atmospheric Measurement Techniques. 2014;**7**:3325-3336

[19] Felix EP, Filho JP, Garcia G, et al. A new fluorescence method for determination of ozone in ambient air. Microchemical Journal. 2011;**99**:530-534

[20] Lee C-M, Kwon M, Kang D-R, et al. Distribution of radon concentrations in child-care facilities in South Korea. Journal of Environmental Radioactivity. 2017;**167**:80-85

[21] Abdel-Salam MMM. Outdoor and indoor factors influencing particulate matter and carbon dioxide levels in naturally ventilated urban homes. Journal of the Air & Waste Management Association (1995). 2021; **71**:60-69

[22] Yip F, Christensen B, Sircar K, et al. Assessment of traditional and improved stove use on household air pollution and personal exposures in rural western Kenya. Environment International. 2017; **99**:185-191

[23] World Health Organization. WHO guidelines for indoor air quality: Selected pollutants. 2010. Available from: http://www.euro.who.int/pubrequest

[24] Kinney PL, Roman HA, Walker KD, et al. On the use of expert judgment to characterize uncertainties in the health benefits of regulatory controls of particulate matter. Environmental Science & Policy. 2010;**13**:434-443

[25] US EPA. IAQ guidelines. 2021. Available from: https://www.epa.gov/

[26] Baldelli A. Evaluation of a low-cost multi-channel monitor for indoor air quality through a novel, low-cost, and reproducible platform. Meas Sensors. 2021;**17**:100059

[27] Chahardoli S, Khakzand M, Faizi M, et al. Numerical analysis of the effect of roof types and porch on particle dispersion and deposition around a low-rise building. Journal of Building Engineering. 2022;**53**:104533

[28] Abanto J, Barrero D, Reggio M, et al. Airflow modelling in a computer room. Building and Environment. 2004;**39**: 1393-1402

[29] Shrestha M, Rijal HB, Kayo G, et al. An investigation on CO_2 concentration based on field survey and simulation in naturally ventilated Nepalese school buildings during summer. Building and Environment. 2022;**207**:108405

[30] Mehdi K, Saeid C, Anahita N, et al. Comparative study of architectural elements to improve the wind environment in hot and humid climates. Journal of Architectural Engineering. 2023;**29**:4023024

[31] Farhadi F, Khakzand M, Altan H, et al. IAQ in CCU units: An experimental and numerical investigation based on the

outlet air height (case study: Namazi Hospital, Shiraz). International Journal of Ventilation. 2023;**0**:1-26

[32] Cooper E, Wang Y, Stamp S, et al. Use of portable air purifiers in homes: Operating behaviour, effect on indoor PM2.5 and perceived indoor air quality. Building and Environment. 2021;**191**: 107621

[33] Settimo G, Avino P. The dichotomy between indoor air quality and energy efficiency in light of the onset of the COVID-19 pandemic. Atmosphere (Basel). 2021;**12**. Epub ahead of print 2021. DOI: 10.3390/atmos12060791

[34] Al-Rawi M. The thermal comfort sweet-spot: A case study in a residential house in Waikato, New Zealand. Case Studies in Thermal Engineering. 2021;**28**: 101530

[35] McArthur JJ, Powell C. Health and wellness in commercial buildings: Systematic review of sustainable building rating systems and alignment with contemporary research. Building and Environment. 2020;**171**:106635

[36] Kim C, Choi D, Lee YG, et al. Diagnosis of indoor air contaminants in a daycare center using a long-term monitoring. Building and Environment. 2021;**204**:108124

[37] Caracci E, Canale L, Buonanno G, et al. Effectiveness of eco-feedback in improving the indoor air quality in residential buildings: Mitigation of the exposure to airborne particles. Building and Environment. 2022;**226**: 109706

[38] Kephart JL, Fandiño-Del-Rio M, Williams KN, et al. Nitrogen dioxide exposures from LPG stoves in a cleaner-cooking intervention trial. Environment International. 2021;**146**:106196

[39] Woolley K, Bartington SE, Pope FD, et al. Biomass cooking carbon monoxide levels in commercial canteens in Kigali, Rwanda. Archives in Environmental Occupational Health. 2021;**76**:75-85

[40] Amadeo B, Robert C, Rondeau V, et al. Impact of close-proximity air pollution on lung function in schoolchildren in the French West Indies. BMC Public Health. 2015;**15**:45

[41] Shen J-H, Wang Y-S, Lin J-P, et al. Improving the indoor air quality of respiratory type of medical facility by zeolite filtering. Journal of the Air & Waste Management Association (1995). 2014;**64**:13-18

[42] Huang L, Mo J, Sundell J, et al. Health risk assessment of inhalation exposure to formaldehyde and benzene in newly remodeled buildings, Beijing. PLoS One. 2013;**8**:e79553

[43] Abdel-Salam MMM. Investigation of PM2.5 and carbon dioxide levels in urban homes. Journal of the Air & Waste Management Association (1995). 2015; **65**:930-936

[44] Poulin P, Leclerc J-M, Dessau J-C, et al. Radon measurement in schools located in three priority investigation areas in the province of Quebec, Canada. Radiational Protocol Dosimetry. 2012; **151**:278-289

[45] Branco PTBS, Alvim-Ferraz MCM, Martins FG, et al. Impact of indoor air pollution in nursery and primary schools on childhood asthma. Science Total Environment. 2020;**745**:140982

[46] Xu H, Zeng W, Guo B, et al. Improved risk communications with a Bayesian multipollutant Air Quality Health Index. Science Total Environment. 2020;**722**:137892

[47] Schmitt LHM. QALY gain and health care resource impacts of air pollution control: A Markov modelling approach. Environmental Science & Policy. 2016; **63**:35-43

[48] Taj T, Jakobsson K, Stroh E, et al. Air pollution is associated with primary health care visits for asthma in Sweden: A case-crossover design with a distributed lag non-linear model. Spatial Spatiotemporal Epidemiology. 2016;**17**: 37-44

[49] Madureira J, Paciência I, Rufo J, et al. Indoor air quality in schools and its relationship with children's respiratory symptoms. Atmospheric Environment. 2015;**118**:145-156

[50] Kanchongkittiphon W, Mendell MJ, Gaffin JM, et al. Indoor environmental exposures and exacerbation of asthma: An update to the 2000 review by the Institute of Medicine. Environmental Health Perspectives. 2015;**123**:6-20

[51] Ming T, Fang W, Peng C, et al. Impacts of traffic tidal flow on pollutant dispersion in a non-uniform Urban Street Canyon. Atmosphere (Basel). 2018;**9**. Epub ahead of print 2018. DOI: 10.3390/atmos9030082

[52] Sicard P, Talbot C, Lesne O, et al. The Aggregate Risk Index: An intuitive tool providing the health risks of air pollution to health care community and public. Atmospheric Environment. 2012; **46**:11-16

[53] Groulx N, Movahhedinia H, Edwards P, et al. Medical air in healthcare institutions: A chemical and biological study. Atmospheric Environment. 2019;**219**:117031

[54] Jung C-C, Wu P-C, Tseng C-H, et al. Indoor air quality varies with ventilation types and working areas in hospitals. Building and Environment. 2015;**85**: 190-195

[55] Nair AN, Anand P, George A, et al. A review of strategies and their effectiveness in reducing indoor airborne transmission and improving indoor air quality. Environmental Research. 2022;**213**:113579

[56] Bartington SE, Bakolis I, Devakumar D, et al. Patterns of domestic exposure to carbon monoxide and particulate matter in households using biomass fuel in Janakpur, Nepal. Environmental Pollution. 2017;**220**:38-45

[57] Peixoto C, Slezakova K, Pereira MDC, et al. Air quality in fitness centers: The impact of ventilation restrictions: A case study. UPorto Journal of Engineering. 2022;**8**:26-35

[58] Agarwal N, Meena CS, Raj BP, et al. Indoor air quality improvement in COVID-19 pandemic: Review. Sustainable Cities and Society. 2021;**70**: 102942

[59] Stamp S, Burman E, Shrubsole C, et al. Long-term, continuous air quality monitoring in a cross-sectional study of three UK non-domestic buildings. Building and Environment. 2020;**180**: 107071

[60] Barcelo D. An environmental and health perspective for COVID-19 outbreak: Meteorology and air quality influence, sewage epidemiology indicator, hospitals disinfection, drug therapies and recommendations. Journal of Environmental Chemical Engineering. 2020;**8**:104006

[61] Kumar P, Morawska L. Could fighting airborne transmission be the next line of defence against COVID-19 spread? City Environmental Interaction. 2019;**4**:100033

[62] Madurai Elavarasan R, Pugazhendhi R. Restructured society and environment: A review on potential technological strategies to control the COVID-19 pandemic. Science Total Environment. 2020;**725**:138858

[63] Megahed NA, Ghoneim EM. Indoor Air Quality: Rethinking rules of building design strategies in post-pandemic architecture. Environmental Research. 2021;**193**:110471

[64] Harrison PTC. Indoor air quality guidelines. Occupational and Environmental Medicine. 2002;**59**:73-74

[65] World Health Organization. WHO guidelines for indoor air quality: Household fuel combustion. 2014. Available from: https://www.who.int/airpollution/guidelines/household-fuel-combustion/IAQ_HHFC_guidelines.pdf

[66] WHO. Guidelines for Indoor Air Quality: Selected Pollutants. Bonn, Germany: Regional Office for Europe; 2010

[67] Al-Hemoud A, Al-Awadi L, Al-Khayat A, et al. Streamlining IAQ guidelines and investigating the effect of door opening/closing on concentrations of VOCs, formaldehyde, and NO_2 in office buildings. Building and Environment. 2018;**137**:127-137

[68] OSHA. Indoor air quality investigation. In: U.S. Department of Labor. Technical manual (OTM). Section III, Chapter 2. Available from: https://www.osha.gov/otm/section-3-health-hazards/chapter-2

[69] Kumar P, Singh AB, Arora T, et al. Critical review on emerging health effects associated with the indoor air quality and its sustainable management. Science Total Environment. 2023;**872**: 162163

[70] Salthammer T. Critical evaluation of approaches in setting indoor air quality guidelines and reference values. Chemosphere. 2011;**82**:1507-1517

[71] Dimitroulopoulou C. Ventilation in European dwellings: A review. Building and Environment. 2012;**47**:109-125

[72] Sun Y, Hou J, Cheng R, et al. Indoor air quality, ventilation and their associations with sick building syndrome in Chinese homes. Energy and Buildings. 2019;**197**:112-119

[73] Yakhot V, Orszag SA, Thangam S, et al. Development of turbulence models for shear flows by a double expansion technique. Physics of Fluids A: Fluid Dynamics. 1992;**4**:1510-1520

[74] Tian L, Ahmadi G. Particle deposition in turbulent duct flows—comparisons of different model predictions. Journal of Aerosol Science. 2007;**38**:377-397

[75] Lecrivain G, Sevan D-M, Thomas B, et al. Numerical simulation of multilayer deposition in an obstructed channel flow. Advanced Powder Technology. 2014;**25**:310-320

[76] Vassella CC, Koch J, Henzi A, et al. From spontaneous to strategic natural window ventilation: Improving indoor air quality in Swiss schools. International Journal of Hygiene and Environmental Health. 2021;**234**:113746

[77] Deng X, Thurston G, Zhang W, et al. Application of data science methods to identify school and home risk factors for asthma and allergy-related symptoms among children in New York. Science Total Environment. 2021;**770**:144746

[78] Ascione F, De Masi RF, Mastellone M, et al. The design of safe classrooms of educational buildings for facing

contagions and transmission of diseases: A novel approach combining audits, calibrated energy models, building performance (BPS) and computational fluid dynamic (CFD) simulations. Energy and Buildings. 2021;**230**:110533

[79] Leconte A, Lafféter C, Fritsch T, et al. Experimental study of the combination of a positive input ventilation and active air vents on the air change rates of a house. International Journal of Ventilation. 2021;**20**:215-225

[80] Cetin Y, Avci M, Aydin O. Effect of air change rate on particle dispersion from inlet opening under varying particle source strengths. International Journal of Ventilation. 2020;**2020**:1-18

[81] Dobson R, O'Donnell R, Tigova O, et al. Measuring for change: A multi-centre pre-post trial of an air quality feedback intervention to promote smoke-free homes. Environment International. 2020;**140**:105738

[82] Lewis JJ, Hollingsworth JW, Chartier RT, et al. Biogas stoves reduce firewood use, household air pollution, and hospital visits in Odisha, India. Environmental Science Technology. 2017;**51**:560-569

[83] Can E, Özden Üzmez Ö, Döğeroğlu T, et al. Indoor air quality assessment in painting and printmaking department of a fine arts faculty building. Atmospheric Pollution Research. 2015;**6**:1035-1045

[84] Baurès E, Blanchard O, Mercier F, et al. Indoor air quality in two French hospitals: Measurement of chemical and microbiological contaminants. Sciene Total Environment. 2018;**642**:168-179

[85] Baboli Z, Neisi N, Babaei AA, et al. On the airborne transmission of SARS-CoV-2 and relationship with indoor conditions at a hospital. Atmospheric Environment. 1994;**2021**(261):118563

[86] Beam EL, Schwedhelm S, Boulter K, et al. Personal protective equipment processes and rationale for the Nebraska Biocontainment Unit during the 2014 activations for Ebola virus disease. American Journal of Infection Control. 2016;**44**:340-342